DIFFERENTIAL EVOLUTION

In Search of Solutions

Optimization and Its Applications

VOLUME 5

Managing Editor
Panos M. Pardalos (University of Florida)

Editor—Combinatorial Optimization
Ding-Zhu Du (University of Texas at Dallas)

Advisory Board
J. Birge (University of Chicago)
C.A. Floudas (Princeton University)
F. Giannessi (University of Pisa)
H.D. Sherali (Virginia Polytechnic and State University)
T. Terlaky (McMaster University)
Y. Ye (Stanford University)

Aims and Scope
Optimization has been expanding in all directions at an astonishing rate during the last few decades. New algorithmic and theoretical techniques have been developed, the diffusion into other disciplines has proceeded at a rapid pace, and our knowledge of all aspects of the field has grown even more profound. At the same time, one of the most striking trends in optimization is the constantly increasing emphasis on the interdisciplinary nature of the field. Optimization has been a basic tool in all areas of applied mathematics, engineering, medicine, economics and other sciences.

The series *Springer Optimization and Its Applications* publishes undergraduate and graduate textbooks, monographs and state-of-the-art expository works that focus on algorithms for solving optimization problems and also study applications involving such problems. Some of the topics covered include nonlinear optimization (convex and nonconvex), network flow problems, stochastic optimization, optimal control, discrete optimization, multi-objective programming, description of software packages, approximation techniques and heuristic approaches.

DIFFERENTIAL EVOLUTION

In Search of Solutions

By

VITALIY FEOKTISTOV

 Springer

Library of Congress Control Number: 2006929851

ISBN-10: 0-387-36895-7   e-ISBN: 0-387-36896-5

ISBN-13: 978-0-387-36895-5   e-ISBN-13: 978-0-387-36896-2

Printed on acid-free paper.

AMS Subject Classifications: 68W01, 68W40, 90C26, 90C56, 90C59, 68T05, 90C30, 65Y20, 65Y05, 65B99, 49Q10

Printed in the United States of America.

9 8 7 6 5 4 3 2 1

springer.com

to my dear teachers and disciples

Contents

1 Differential Evolution 1
   1.1 What Is Differential Evolution? 1
   1.2 Its History and Development 2
   1.3 What Purpose Does It Serve? 7
   1.4 The Famous Algorithm 13
   1.5 Secrets of a Great Success 17
   Problems 21

2 Neoteric Differential Evolution 25
   2.1 Evolutionary Algorithms 25
   2.2 Problem Definition 28
   2.3 Neoteric Differential Evolution 28
   2.4 Distinctions and Advantages 30
   2.5 Mixed Variables 32
   2.6 Constraints 33
      2.6.1 Boundary Constraints 33
      2.6.2 Constraint Functions 34
   Problems 37

3 Strategies of Search 41
   3.1 Antecedent Strategies 41
   3.2 Four Groups of Strategies 42
      3.2.1 RAND Group 44
      3.2.2 RAND/DIR Group 44
      3.2.3 RAND/BEST Group 45
      3.2.4 RAND/BEST/DIR Group 46
      3.2.5 On the Constant of Differentiation 46
   3.3 Examples of Strategies 47
      3.3.1 RAND Strategies 47
      3.3.2 RAND/DIR Strategies 51
      3.3.3 RAND/BEST Strategies 56

  3.3.4   RAND/BEST/DIR Strategies 58
3.4 Tests 63
Problems 66

4 Exploration and Exploitation 69
4.1 Differentiation via Mutation 70
4.2 Crossover 70
4.3 Analysis of Differentiation 71
4.4 Control Parameters 73
  4.4.1   Diversity Estimation 73
  4.4.2   Influence of Control Parameters 75
  4.4.3   Tuning of Control Parameters 76
4.5 On Convergence Increasing 77
Problems 79

5 New Performance Measures 83
5.1 Quality Measure (*Q-Measure*) 83
5.2 Entropy 85
5.3 Robustness Measure (*R-Measure*) 86
5.4 Population Convergence (*P-Measure*) 86
Problems 89

6 Transversal Differential Evolution 91
6.1 Species of Differential Evolution 91
6.2 Two-Array Differential Evolution 92
6.3 Sequential Differential Evolution 93
6.4 Transversal Differential Evolution 94
6.5 Experimental Investigations 95
6.6 Heterogeneous Networks of Computers 97
Problems 100

7 On Analogy with Some Other Algorithms 101
7.1 Nonlinear Simplex 102
7.2 Particle Swarm Optimization 104
7.3 Free Search 106
Problems 109

8 Energetic Selection Principle 111
8.1 Energetic Approach 111
8.2 Energetic Selection Principle 113
  8.2.1   Idea 113
  8.2.2   Energetic Barriers 113
  8.2.3   Advantages 116
8.3 Comparison of Results 117
Problems 119

9   On Hybridization of Differential Evolution 121
   9.1   Support Vector Machine 121
   9.2   Hybridization 123
   9.3   Comparison of Results 125
   9.4   Some Observations 129
   Problems 132

10   Applications 133
   10.1   Decision Making with Differential Evolution 133
      10.1.1   Aggregation by the Choquet Integral 134
      10.1.2   Classical Identification Problem 135
      10.1.3   Implementation and Comparison of Results 136
   10.2   Engineering Design with Differential Evolution 139
      10.2.1   Bump Problem 139
      10.2.2   The Best-Known Solutions 139
      10.2.3   Implementation and Comparison of Results 141

11   End Notes 145

A   Famous Differential Evolution 149
   A.1   C Source Code 149
   A.2   MATLAB Source Code 153

B   Intelligent Selection Rules 157

C   Standard Test Suite 159
   C.1   Sphere Function 159
   C.2   Rosenbrock's Function 159
   C.3   Step Function 160
   C.4   Quartic Function 162
   C.5   Shekel's Function 162
   C.6   Rastrigin's Function 163
   C.7   Ackley's Function 163
   C.8   Rotated Ellipsoid Function 164

D   Coordinate Rotation of a Test Function 167

E   Practical Guidelines to Application 171

References 173

Index 187

Preface

Differential evolution is one of the most recent global optimizers. Discovered in 1995 it rapidly proved its practical efficiency. This book gives you a chance to learn all about differential evolution. On reading it you will be able to profitably apply this reliable method to problems in your field.

As for me, my passion for intelligent systems and optimization began as far back as during my studies at Moscow State Technical University of Bauman, the best engineering school in Russia. At that time, I was gathering material for my future thesis. Being interested in my work, the Mining School of Paris proposed that I write a dissertation in France. I hesitated some time over a choice, but my natural curiosity and taste for novelty finally prevailed. At present, *Docteur ès science en informatique de l'École des Mines de Paris*, I am concentrating on the development of my own enterprise. If optimization is my vocation, my hobbies are mathematics and music. Although mathematics disciplines the mind, music is filled with emotions. While playing my favorite composition, I decided to write this book.

The purpose of the book is to give, in a condensed but overview form, a description of differential evolution. In addition, this book makes accessible to a wide audience the fruits of my long research in optimization. Namely, I laid the foundation of the universal concept of search strategies design, suitable not only for differential evolution but for many other algorithms. Also, I introduced a principle of energetic selection, an efficient method of hybridization, and advanced paralleling techniques.

In spite of the scientific character, this book is easy to read. I have reduced the use of mathematical tools to a minimum. An understanding of college-level mathematics (algebra and a little calculus) is quite enough. The book is designed for students, teachers, engineers, researchers, and simply amateurs from very different fields (computer science, applied mathematics, physics, engineering design, mechanical engineering, electrical engineering, bioinformatics, computational chemistry, scheduling, decision making, financial mathematics, image processing, neuro-fuzzy systems, biosystems, and control theory).

This material may be used as a basis for a series of lectures. For an introductory course you need only Chapter 1 and Sections 2.1–2.2. For an advanced course you may use Chapters 2–6. For those interested in profound study and new ideas in optimization I suggest reading Chapters 7–9 too. Those who are searching for practical applications may read only Section 1.2 and Chapter 10.

Perhaps you want to be an extraordinaire and begin reading this book from the end notes (Chapter 11). Good idea! Enjoy!

Acknowledgments

In the first place I thank Kenneth Price, without whom differential evolution would not exist, for our fruitful correspondence.

Many of my friends and colleagues have also helped by commenting on drafts, discussing ideas, and answering questions. I thank in particular Stefan Janaqi and Jacky Montmain.

I express profound gratitude to the Technological Incubator of EMA, especially to Michel Artigue, Laetitia Leonard, and Yannick Vimont for a keen appreciation of my project.

I thank everyone at Springer warmly, especially Angela Burke and Robert Saley, who have been helpful throughout.

All my sincere appreciation is directed to Olga Perveeva, my best friend. I am grateful to you for your permanent help.

I tender thanks to my parents, Lubov and Andrey Feoktistov, for their warm encouragement and moral support.

Special thanks to my Russian–English and English–Russian dictionaries as well as my English grammar book; without them writing this treatise would have been very arduous.

Nîmes, France,   *Vitaliy Feoktistov*

1

Differential Evolution

In this chapter I explain what differential evolution is. We speak about
its history and development as well as the purposes it serves. Moreover,
I immediately introduce a famous version of the differential evolution
algorithm and clarify why and how it works.

1.1 What Is Differential Evolution?

Even without thinking, steadily or almost magically, day by day we optimize
our everyday life. We explore new areas, make decisions, do something, or
maybe do nothing and are simply relaxed and having a rest. Continually,
we pass from one state to another and involuntarily feel as if something new
comes into us. We agree and believe furthermore that we are on the right track
and have just achieved something important and essential. In other words, we
have truly evolved to a new stage. So, we speak about **evolution**.

Many routes, many choices. How do we choose our own trip? Almost every-
body will agree with me in that there is something determinative. For someone
it will be a comfortable, rich, and stable life; for someone else it will be a life
full of adventures; for others both things are important. An **individual** may
be inclined to rational behavior or may completely rely on his or her sixth
sense. Many of us permanently trust to intuition. This is a magic sense that
incessantly guides us through life helping us make a right choice, of course,
if we want it. All this spontaneously suggests a **criterion** of choice. And we
choose.

I am not the one and only in this world. I am generously circled by interesting people, spectacular events, and relations. All together we form a society
or, biologically speaking, a **population**. From this point onwards, in addition
to individual peculiarities, social behavior is automatically included. Perhaps
you have already heard that one of most influential features of social activity
is a **collective intelligence**. The main principle here is an integration, the

integration of different lives, different minds, and different states in a single whole in order to be more powerful, more efficient, and more intelligent, and the more the better. Lots of scientists from different fields of science repeatedly described and will describe such a phenomenon.

> **Differential evolution is a small and simple mathematical model of a big and naturally complex process of evolution. So, it is easy and efficient!**

First and foremost Differential Evolution (DE) is an optimization algorithm. And without regard to its simplicity DE was and is one of the most powerful tools for **global optimization**.

Perhaps you have heard about genetic algorithms, evolution strategies, or evolutionary programming. These are three basic trends of evolutionary optimization, also well known under the common term of **evolutionary algorithms**. Lately, with the advent of new ideas and new methods in optimization, including DE too, they have got a second fashionable name of **artificial evolution**. DE belongs to this suite.

As you are already guessing, DE incarnates all the elements described above; namely it realizes the *evolution of a population of individuals in some intelligent manner*. "How does DE do it?" and "What ideas are behind it?" is the theme of the book and I omit them now. However, it is very likely that you want to know right now what the difference is between DE and the other algorithms. Without going into detail, I outline it.

> **Intelligent use of differences between individuals realized in a simple and fast linear operator, so-called differentiation, makes differential evolution unique.**

Now you may presume why DE was called *differential* evolution. Yes, of course, you are right! And that skillful manipulation of individual differences plays a paramount role. Also, I always bear in mind the origins of DE. For example, as you can sense from the names, well-known genetic algorithms spring up from the genetic evolution of chromosomes, ant colony optimization is guided by the study of ants' behavior, particle swarm optimization is based on the study of social phenomena, and so on. In contrast to all, DE was derived from naturally mathematical (geometrical) arguments. And just this strict mathematics permits a pure and pragmatic exploitation of available information without any notional restrictions. It is easy and efficient!

1.2 Its History and Development

And now, a little bit of history. Also, I am going to speak you about different axes of development. For somebody who is not familiar enough yet with evolutionary optimization, it is possible that not all the terms of this section will be absolutely clear. Don't worry, skip the obscure term and go on reading. In the following chapters we refer to it in detail. This story is destined for the curious, those who want to know all about differential evolution right now from the first lines. So, ...

Genetic annealing developed by K. Price [Pri94] was the beginning of the DE algorithm. The first paper about genetic annealing was published in the October 1994 issue of *Dr. Dobb's Journal*. It was a population-based combinatorial algorithm that realized an annealing criterion via thresholds driven by the average performance of the population. Soon after this development, Price was contacted by R. Storn, who was interested in solving the Tchebychev polynomial fitting problem by genetic annealing. After some experiments Price modified the algorithm using floating-point instead of bit-string encoding and arithmetic vector operations instead of logical ones. These recasts have changed genetic annealing from a combinatorial into a **continuous optimizer**.

In this way, Price discovered the procedure of **differential mutation**. Price and Storn detected that differential mutation combined with discrete recombination and pairwise selection does not need an annealing factor. So, the annealing mechanism had been finally removed and thus the obtained algorithm started the era of differential evolution.

For the first time differential evolution was described by Price and Storn in the ICSI technical report ("Differential evolution — A simple and efficient adaptive scheme for global optimization over continuous spaces", 1995) [SP95]. One year later, the success of DE was demonstrated in May of 1996 at the First International Contest on Evolutionary Optimization, which was held in conjunction with the 1996 IEEE International Conference on Evolutionary Computation [SP96]. The algorithm won third place for proposed benchmarks.

Inspired by the results, Price and Storn wrote an article for *Dr. Dobb's Journal* ("Differential Evolution: A simple evolution strategy for fast optimization") which was published in April 1997 [PS97]. Also, their article for the *Journal of Global Optimization* ("Differential evolution — A simple and efficient heuristic for global optimization over continuous spaces") was soon published, in December 1997 [SP97]. These papers introduced DE to a large international public and demonstrated the advantages of DE over the other heuristics. Very good results had been shown on a wide variety of benchmarks.

Furthermore, Price presented DE at the Second International Contest on Evolutionary Optimization in 1997 ("Differential evolution vs. the functions of the second ICEO") [Pri97]. There, DE was one of the best among emulous algorithms. And finally, two years later, in 1999, he summarized the algorithm in the compendium "New Ideas in Optimization" [Pri99].

Also, other papers of Rainer Storn can be enumerated here [Sto95, Sto96b, Sto96a, Sto99]. He had been concentrating on various DE applications and had published his Web site (http://www.icsi.berkeley.edu/~storn/code/) containing source codes and many useful links.

In 1998 J. Lampinen set up the official bibliography site [Lam02a] (`http://www.lut.fi/~jlampine/debiblio.html`), which furnishes all materials and also some links on DE dated from 1995 up to 2002.

Right here the history is coming to the end, and we immediately proceed to the development.

As stated in [SP95], the key element distinguishing DE from other population-based techniques is differential mutation. The initial set of **strategies** realizing differential mutation was proposed by Storn and Price in [Sto96a, SP95] and `http://www.icsi.berkeley.edu/~storn/de36.c`. The first attempt to guide differential mutation was presented by Price in [Pri97], where "semi-directed" mutation was realized by a special preselection operation. Later, in [Pri99], Price analyzed the strategies and noted that the strategy may consist of differential mutation and arithmetic crossover. This, in turn, gives the different dynamic effects of search.

The ideas of "directions" were spontaneously grasped by H.-Y. Fan and J. Lampinen. In 2001, they proposed alternations of the classical strategy (the first strategy suggested by Price) with a triangle mutation scheme [FL01] and, in 2003, alternations with a weighted directed strategy, where they used two difference vectors [FL03]. These methods give some improvements, but it is also necessary to note that the percentage of using novel strategies is quite moderate.

In my research I continued the development of strategies and proposed the unique conception of a strategy construction. In 2004, this conception was demonstrated by the example of a group of directed strategies [FJ04g]. From this point on, there is a unique formula that describes all the strategies and clearly reflects the fundamental principle of DE. The strategies were divided into four groups [FJ04d]. Each of the groups was associated with a certain type/behavior of search: random, directed, local, and hybrid. Thorough investigations and test results of some strategies were published in [FJ04b]. In addition, I suggested in [FJ04c] a combinatorial approach to estimate the potential diversity of a strategy. This approach contributed to the correct strategy's choice. The operation realizing a strategy in the DE algorithm was called *differentiation*.

Let's now consider a **crossover** operation. For DE two types of combinatorial crossovers were implemented: binary and exponential ones [Pri99]. The superiority of each crossover over the other cannot be uniquely defined. As for a **selection** operation, the pairwise selection, also so-called "greedy" selection or elitist selection, is steadily used in the algorithm.

The next stage was the introduction of **mixed variables**. In 1999, I. Zelinka and J. Lampinen described a simple and, at the same time, efficient way of handling simultaneously continuous, integer, and discrete variables [LZ99b]. They applied this method to design engineering problems [LZ99a, Lam99]. The obtained results outperformed all the other mixed-variables methods used in engineering design [LZ99c]. As a particular case of mixed-variable problems, in 2003, I implemented DE in the binary-continuous large-scale application in the frame of the ROADEF2003 challenge [FJ03].

Let me now pass to **constraints**. In order to handle boundary constraints two solutions can be implemented: (1) reinitialization [SP95] and (2) periodic mode (or shifting mechanism) [ZX03, MCTM04]. For other constraints (mostly nonlinear functions) penalty methods are used [LZ99b, Mic97] as well as the modification of selection rules [FF97, Deb00], first reported for DE, in 2001, by Lampinen [Lam01, Lam02b] and later, in 2004, by Coello Coello et al. [MCTM04]. The comprehensive bibliography of constraint methods for evolutionary optimization can be found on the following site: http://www.cs.cinvestav.mx/~constraint/.

The question of an algorithm architecture had been untouched for many years. Since the birth of DE, two-array **species** were generally accepted [SP95], justified by their easy parallelization. However, personally I [FJ04b] and the other DE researchers naturally prefer sequential species intuitively believing in their superiority. In 2004, I revealed this question in the comparative study of DE species. It led me to discover an intermediate species: *transversal* DE [FJ04h]. From here, some population-based optimizers (e.g., particle swarm optimization (PSO) [Ang98] and free search (FS) [Pen04]) can be easily interpreted by analogy with transversal DE. Moreover, this species is well adapted for parallelization on heterogeneous networks of computers [Las03]. Thus, DE has been attaining perfection.

The next important point is **control parameters**. DE disposes three control parameters: population size, differentiation constant, and crossover constant. In spite of the fact that DE is more robust regarding control parameters in comparison with, for example, particle swarm optimization or evolutionary algorithms [VT04], nevertheless a well-chosen set of control parameters improves the algorithm's convergence considerably.

At first, there were some recommendations on how to choose an appropriate control parameter set: [SP95, Sto96a, SP96, PS97, Pri97, SP97, Pri99, LZ00]. Then, being influenced by Beyer's [Bey98] postulate,

> the ability of an EA to find the global optimal solution depends on its ability to find a right relation between exploitation of elements found so far and exploration of the search space,

many scientists tried to estimate the diversity of population. Here were proposed: expected population variance [Zah01], average population diversity [Š02], mean square diversity [LL02a], and P-measure [FJ04b, FJ04c]. At once, there were realized adaptation schemes to control the desired diversity level: decreasing of population size [FJ04f], refreshing of population [Š02], use of precalculated differentials [AT02], adaptation of differentiation constant [Zah02], its fuzzy control [LL02a], and relaxation [RL03, AT02, MM05] as well as self-adaptation of the differentiation and crossover constants [Abb02] and strategies [QS05]. Also, the analysis of a stagnation effect was made [LZ00].

Another way to increase convergence is **hybridization**. Hybridization can be settled at four levels of interaction:

1. *Individual* level or search space exploration level, describes the behavior of an individual in the population.
2. *Population* level, represents the dynamics of a population or subpopulations.
3. *External* level, provides the interaction with other methods.
4. *Meta* level, at this level a superior metaheuristics includes the algorithm as one of its strategies.

On the individual level: There were attempts, made in 2003 by Zhang and Xie, to combine PSO with classical DE [ZX03]. Also, T. Hendtlass tried to alternate PSO and DE [Hen]. On the population level: I proposed to add an energetic filter for the first generations in order to eliminate "bad" individuals from the population [FJ04f]. On the external level: I suggested an extra, SVM-based, function to approximate the optimal or "good" individuals using the current population state [FJ04e]. Also, there are a series of works on large-scale optimization [AT00, AT02], where DE is hybridized with the L-BFGS algorithm of Lui and Nocedal [LN89]. Here, a topographical principle is used to define the "stars" for local optimization. At last, on the meta level: the integration of DE into a set of competing heuristics was shown by J. Tvirdik in [Tvi04]. There, heuristics are used at random with a certain probability that depends on the success of heuristics in the preceding steps. Furthermore, Xie and Zhang tend to develop agent-based modeling for solving optimization problems by swarm algorithms. In their case, DE is presented as one of the "generate-and-test" rules [XZ04].

In addition to the above, DE was enlarged on both **multimodal** and **multiobjective optimization**. There are many domains where searching multiple global optima is an important task (e.g., power system stability, digital filter design, electromagnetics, DNA sequence analysis, etc.). In the work of J. Rumpler and F. Moore [RM01] DE was modified so that it is capable of automatically determining the needed number of subpopulations as well as the minimal spanning distance between individuals of each subpopulation. Later, in 2004, R. Thomsen introduced crowding-based DE [Tho04]. Such a variation outperforms the well-known sharing scheme that penalizes similar candidate solutions. Also in 2004, D. Zaharie illustrated a parallel implementation of DE for multipopulation optimization. The main purposes were to find all possible, global and/or local, optima and to speed up the algorithm [Zah04]. Furthermore, she studies the migration concept between subpopulations: neither niche radius nor global clustering are needed in this case.

In the papers of H. Abbass, R. Sarker, and C. Newton [ASN01a, ASN01b, AS02, SA04] and independently in the paper of N. Madavan [Mad02] the ideas of the DE pareto-approach for multiobjective optimization were developed. The solutions provided by the proposed algorithm outperform the Strength Pareto Evolutionary Algorithm, one of the best evolutionary algorithms for multi-criterion optimization. Then, H. Abbass introduced a self-adaptation for both the crossover and differentiation constants [Abb02]. It made the algorithm still more competitive. The next interesting point is that H. Abbass together with K. Deb analyzed a multiobjective approach applied to a single objective function [AD03]. The experiments showed that such a technique increases both the convergence rate and precision of a solution. Simultaneously, S. Kukkonen and J. Lampinen examined the constrained multiobjective DE algorithm applied to engineering problems [KL04, KSL04, KL05]. And D. Zaharie spread the migration concept on pareto-optimization [ZP03, Zah03]. There also exists a promising multiobjective DE modification, developed by a group of Greek researchers, based on the multipopulation concept [PTP⁺04]. Finally, continuous and discrete multiobjective DE (C-MODE and D-MODE, respectively) are perfectly summarized by American researchers in [XSG03b, Xue04, XSG05b, XSG05a].

Now, in conclusion of this section, we can truly evaluate the great power and importance of differential evolution. Below, in Fig. 1.1, I have represented a summarizing scheme. This scheme synthesizes the state of the art of the algorithm in a graphical form and, if you don't mind, I would venture to emphasize in it my personal contribution to some of its aspects.

1.3 What Purpose Does It Serve?

What purpose does it serve, differential evolution? Of course, as you already know, for optimization. But what kind of optimization? Or again, for what type of real problems is differential evolution better suited? In order to be obvious, I propose that you examine the following short scheme of optimization methods (see Fig. 1.2).

As we can see from this scheme, it is customary to divide all optimization methods into two large classes: (1) continuous optimization, where the search area and solutions are presumed to be situated in a certain continuous space with its metrics; and (2) combinatorial optimization, where the search area is limited by a finite number of feasible solutions. Depending on what type of function a real problem is formulated into, continuous optimization methods are subdivided, broadly speaking, into linear programming (linear objective function and linear constraints), quadratic programming (quadratic objective function and linear constraints), and nonlinear programming (nonlinear objective function and nonlinear constraints). In nonlinear programming local search methods are the most used. Often, in classical methods, global search is successfully realized by solving a sequence of local optimization problems. As for combinatorial methods, they are mainly subdivided into two categories.

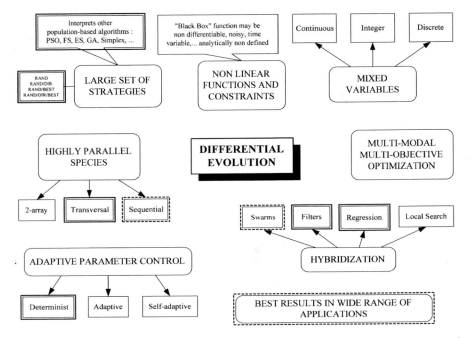

Fig. 1.1. The state of the art of differential evolution. Second solid line: my total contribution to DE; Second dotted line: my partial contribution to DE.

The first category is exact methods, where the global optimum is attained by enumerating all sets of solutions. This approach is time consuming and is good only for small-scale problems. The second category is approximate methods, where a partial enumeration is applied to attain a near-to-optimum solution, which represents a time/quality compromise. The most commonly used here are heuristic methods, that are usually designed specially for a certain problem. More often than not they are not flexible enough to be applied to another problem.

However, in some real situations it is more natural to model a problem by mixed variables; that is, one part of such a model contains variables that are allowed to vary continuously and in the other part variables can attain only discrete values. The well-known special case is mixed-integer (linear, quadratic, nonlinear) programming, where the discrete sets consist only of integer values.

There is a class of problems that can be modeled by continuous, combinatorial, or even mixed variables, where we are not able to find the global solution by applying traditional methods. I mean here by traditional methods the classical, or determinist, methods in continuous optimization [NW99] and the heuristics in combinatorial optimization [NW88, PS82]. So, in the con-

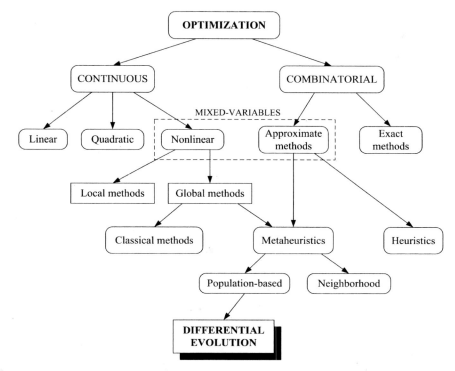

Fig. 1.2. A simple classification scheme of optimization methods: position of differential evolution.

tinuous case, one does not know the algorithm allowing us to find the global optimum. And in the combinatorial case, one presumes the nonexistence of the polynomial optimizer [PS82]. This class of problems was called problems of *difficult optimization* [DPST03]. At the present time, these problems are the most attractive and most urgent in the optimization domain.

During the last few years *metaheuristic methods* have become more and more available for difficult optimization [PR02]. They are more universal and less exacting with respect to an optimization problem. It is for this reason that metaheuristics are getting the last word in fashion. As their advantages we can accentuate the following facts.

- They do not require special conditions for the proprieties of the objective function and constraints.
- They can be applied for both continuous problems and combinatorial ones.
- They are extensible on multimodal and multiobjective optimization.

In contrast, of course, there are some disadvantages:

- Lack of strong convergence proof

- Sensitivity of the control parameters
- Computing time can be high enough

It is customary to classify all the metaheuristics into two groups:

1. *Population-based* metaheuristics, where a set of solutions simultaneously progresses towards the optimum. As examples of this case genetic algorithms [Gol89], particle swarm optimization [KE95], and also differential evolution can be cited; and
2. *Neighborhood* metaheuristics; here only one solution is advanced time and again. In this case simulated annealing [KGV83, vA92] and tabu search [Glo90] are universally known examples.

The group of *population-based* metaheuristics intrinsically possesses some important advantages, namely:

- They provide information concerning the "surface" of an objective function.
- They are less sensitive to "improper" pathways of certain individual solutions.
- They increase the probability of attaining the global optimum.

> **Differential evolution, being a population-based optimizer, has made a high-grade break-through.**

As we can see from Fig. 1.2, differential evolution, deriving from population-based metaheuristics, inherits all the best properties of its ancestors: global methods of nonlinear continuous optimization, approximate methods of combinatorial optimization, mixed-variables handling, and so on. In addition, it provides stochastic optimization principles, distributed searchers, and universal heuristics. All this makes the algorithm a general-purpose optimizer, remaining, with all this going on, simple, reliable, and fast.

So to back up my statements I would like to present a simple example that demonstrates some characteristics of the population-based approach. I'll take the well-known two-dimensional Rosenbrock's (or banana) function:

$$f(x, y) = 100 \cdot (x^2 - y)^2 + (1 - x)^2 . \tag{1.1}$$

It is called the "banana" function because of the way the curvature bends around the origin. It is notorious in optimization examples because of the slow convergence that most methods exhibit when trying to solve this problem. This function has a unique minimum at the point $[1, 1]$, where $f(x, y) = 0$.

Firstly, we apply steepest descent as proposed in MATLAB. Starting at the point $[-1.9, 2]$ and after 301 function evaluations it finally stagnates far from the global optimum (see Fig. 1.3, left).

Then, we use differential evolution to solve the same problem. Initializing the population within boundary constraints $[-2.048, 2.048]$ and using a standard control parameter set, after 10 iterations we can see how the population

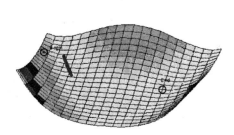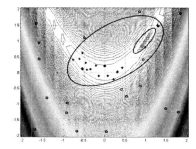

Fig. 1.3. Steepest descent (left) and differential evolution (right) applied to the banana function.

approaches to the optimal zone (see the big ellipse in Fig. 1.3, right) and after only 20 iterations (400 function evaluations) the population occupies the zone just around the optimum (see the small ellipse in Fig. 1.3, right). The best individual of the population already takes precise values $x = 1.00000000515790$ and $y = 1.00000001159138$, which gives $f(x,y) = 1.893137459\text{e}{-}16 \approx 0$.

Thus, for population-based algorithms, and in particular for differential evolution, we can observe the following.

1. Global optimum is attained.
2. Excellent precision.
3. Fast convergence.
4. Self-adaptation.
5. 0-order information about the objective function.

Next, I'll change the steepest descent method to a Broyden–Fletcher–Goldfarb–Shanno (BFGS) one. In this case the optimum on the banana function can be found after 85 function evaluations (faster than DE). But, if we want to use the BFGS algorithm for multimodal optimization, for instance, if I take Rastrigin's function (see Fig. 1.4), it will stagnate at the first local minimum, whereas DE successfully finds the optimum $f(0,0) = 0$ after 1000 function evaluations with precision less than $1.0\text{e}{-}6$. Below I have presented a 2-D exemple. Also it should be noted that the characteristics of the population-based approach remain steadily true and for higher dimensions.

As you already understood, differential evolution is so universal that it can be applied to practically any optimization problem, whether it is linear/nonlinear, continuous or combinatorial, or else a mixed-variable one. Although,

> **Differential evolution was originally mean for difficult optimization on continuous space and is best in those cases where traditional methods come off second best.**

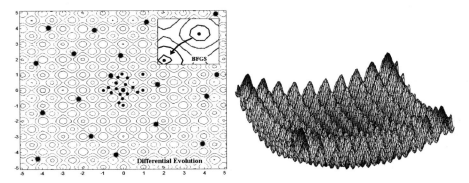

Fig. 1.4. BFGS and differential evolution applied to Rastrigin's function. Left: contour lines of the function, both methods BFGS and DE are shown; right: 3-D view of the function.

The best way to show the importance of the method is to outline the circle of applications in which this method can be successfully implemented. Thus, in conclusion of this section I shall enumerate families (see Fig. 1.5) and give some references to the most significant **applications** of Differential Evolution.

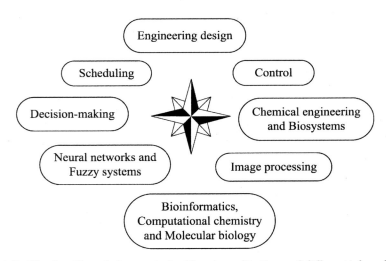

Fig. 1.5. The families of the most significant applications of differential evolution.

Engineering design: Parameter filter design [Sto95, Sto96b], mechanical engineering design [LZ99c, LZ99a], aerodynamic design [Rog98, RDK99a,

RDK99b, RDK99c, RKD00, RD00], radio-frequency integrated circuit design [VRSG00].

Control: Optimal control and optimal time location problems [WC97b, WC97a], fuzzy controller design [CL99] and fuzzy controller tuning [CXQ99, CX00], multimodal optimal control problems [CWS01], parameter identification [UV03], path planning optimization [SMdMOC03].

Scheduling: Scheduling in parallel machine shop [Rüt97a, Rüt97b], plant scheduling and planning [LHW00], management of the mission of Earth observation satellites [FJ03], enterprise planning [XSG03a], heterogeneous multiprocessor scheduling [RS05].

Chemical engineering and biosystems: Fermentation processes [CW99], optimal control of bioprocess systems [CW98].

Bioinformatics, computational chemistry, and molecular biology: Flexible ligand docking problem [Tho03], potential energy minimization [AT00, MA03].

Neural networks and fuzzy systems: Training for feedforward networks [IKL03], training and encoding into feedforward networks [wCzCzC02], online training [MPV01], tuning fuzzy membership functions [MBST99].

Decision making: Fuzzy approach using mixed-integer hybrid DE [HW02], identification by Choquet Integral [FJ04a].

Image processing: Image registration [TV97, Sal00, SPH00], unsupervised image classification [OES05].

1.4 The Famous Algorithm

Now I shall demonstrate the famous DE algorithm, the algorithm that was first proposed by Kenneth Price and Rainer Storn. Namely, this algorithm began the era of differential evolution and certainly thanks to this algorithm differential evolution has been gaining world wide prestige.

The purpose of this section is to state on paper a simple version of DE so that after reading this you would be able to create your own differential evolution using the programming language you prefer.

Let us assume that we are searching the optimum solution X^*, which is presented by a vector of parameters x_i^*, $i = 1, \ldots, D$, subject to boundary constraints $L \leq X \leq H$. The criterion of optimization is reflected by some scalar function $f(X)$ that we want, for example, to minimize. In other words we state the following optimization problem

$$\min_{X} f(X), \quad L \leq X \leq H, \quad X \in \mathbb{R}^D.$$

Before optimizing we should prepare a random number generator to generate the uniform sequence of floating-point numbers in the $[0, 1)$ range. I like

MT19937f due to the fact that this generator has excellent statistical qualities, a very long period, and an impressive speed. Set $rand_i[0, 1)$ as the ith random number produced by the generator.

Then, we need a *test function*. I have used the same functions (1.1) shown in Fig.1.3, Rosenbrock's function, but generalized for higher dimensions:

$$f(X) = \sum_{i=1}^{D-1} 100 \cdot (x_i^2 - x_{i+1})^2 + (1 - x_i)^2, \qquad -2.048 \leq x_i \leq 2.048.$$

You may use your favorite test suite instead of this function for better appreciation of the algorithm.

Of course, we must also tune several optimization parameters. For convenience I have joined together all needed parameters under the common name *control parameters*, although, as a matter of fact, there are only three real control parameters in the algorithm. These are: (1) differentiation (or mutation) constant F, (2) crossover constant Cr, and (3) size of population NP. The rest of the parameters are (a) dimension of problem D that scales the difficulty of the optimization task; (b) maximal number of generations (or iterations) GEN, which serves as a stopping condition in our case; and (c) low and high boundary constraints, L and H, respectively, that limit the feasible area. You can vary all these parameters at will. I set their values to some middle position to demonstrate the average behavior of the algorithm: $D = 10$, $NP = 60$, $F = 0.9$, $Cr = 0.5$, $GEN = 10000$, $L = -2.048$, and $H = 2.048$.

Let me now declare operating variables: $X \in \mathbb{R}^D$, trial individual (or solution); $Pop \in \mathbb{R}^{[D \times NP]}$, population, which represents a set of potential solutions; $Fit \in \mathbb{R}^{NP}$, fitness of the population; $f \in \mathbb{R}$, fitness of the trial individual; $iBest \in \mathcal{N}$, index of the current best individual in the population; $i, j, g \in \mathcal{N}$, loop variables; $Rnd \in \mathcal{N}$, mutation parameter; and $r \in \mathcal{N}^3$, indices of randomly selected individuals.

Before launching the optimization we have to initialize the population and evaluate its fitness (criterion of optimization). Always supposing that we don't know the information about the optimum, the population is initialized within the boundary constraints $Pop_{ij} = L + (H - L) \cdot rand_{ij}[0, 1)$, $i = 1, \ldots, D$, $j = 1, \ldots, NP$. So, the fitness of population is $Fit_j = f(Pop_j)$.

From here on I shall supplement some selective parts of the algorithm with its source code. For that I utilize the C language as the most extensively used in practice. The complete code of this algorithm, written in C and MATLAB, can be found in Appendix A. Thus, the initialization is implemented in the following way.

```
for (j=0; j<NP; j++)
{
    for (i=0; i<D; i++)
        Pop[i][j] = X[i] = L + (H-L) * rand() ;
```

```
    Fit[j] = fnc(D,X) ;
}
```

Now we are ready for optimization. The algorithm is iterative. It makes GEN iterations, where GEN is used as a stopping condition. At each iteration, for each individual the next four steps are realized:

1. Random choice of three individuals from the population, mutually different and also different from the current individual j
   $r_{1,2,3} \in [1, \dots, NP]$, $r_1 \neq r_2 \neq r_3 \neq j$

```
    r[0] = (int) (rand()*NP) ;
    while (r[0]==j)
        r[0] = (int) (rand()*NP) ;
    r[1] = (int) rand()*NP ;
    while ((r[1]==r[0])||(r[1]==j))
        r[1] = (int) (rand()*NP) ;
    r[2] = (int) (rand()*NP) ;
    while ((r[2]==r[1])||(r[2]==r[0])||(r[2]==j))
        r[2] = (int) (rand()*NP) ;
```

2. Creation of the trial individual X. For the first, the mutation parameter Rnd is randomly selected from the $[1, \dots, D]$ range. It guarantees that at least one of parameters will be changed. Then, the trial individual is constructed according to the next probabilistic rule

$$x_i = \begin{cases} x_{i,r_3} + F \cdot (x_{i,r_1} - x_{i,r_2}) & \text{if } (rand_{ij}[0,1) < Cr) \vee (Rnd = i) \\ x_{ij} & \text{otherwise} \end{cases}$$
$$i = 1, \dots, D$$

$$(1.2)$$

```
    Rnd = (int)(rand()*D) ;
    for (i=0; i<D; i++)
    {
        if ( (rand()<CR) || (Rnd == i) )
            X[i] = Pop[i][r[2]] +
                F * (Pop[i][r[0]] - Pop[i][r[1]]) ;
        else
            X[i] = Pop[i][j] ;
    }
```

3. Verifying the boundary constraints. If some of parameters of the trial individual violate the constraints, it is naturally returned into the feasible area.

$$\text{if} \quad (x_i \notin [L, H]) \qquad x_i = L + (H - L) \cdot rand_i[0,1)$$

```
for (i=0; i<D; i++)
    if ((X[i]<L)||(X[i]>H))
        X[i] = L + (H-L) * rand() ;
```

4. Selection of the best individual. First the fitness functions of the trial and current individuals are compared. If the trial's function is less than or equal to the current one, the trial individual replaces the current individual in the population. Besides, it is necessary to check whether the new member of the population is better than the instant best individual. If this is the case, the best individual's index is updated as well.

$$\text{if} \quad (f(X) \leq Fit_j)$$
$$Pop_j \leftarrow X, \ Fit_j \leftarrow f(X) \ \text{and if} \ \ (f(X) \leq Fit_{iBest}) \quad iBest \leftarrow j$$

```
f = fnc(D,X) ;
if (f <= Fit[j])
{
    for (i=0; i<D; i++)
        Pop[i][j] = X[i] ;
    Fit[j] = f ;

    if (f <= Fit[iBest])
        iBest = j ;
}
```

When the algorithm is finally completed, the optimal solution is Pop_{iBest} with its fitness value Fit_{iBest}. I want you to bear this algorithm in mind, so I shall summarize it in a few lines.

Algorithm 1 Famous Differential Evolution

Require: D – problem dimension (optional)
        NP, F, Cr – control parameters
        GEN – stopping condition
        L, H – boundary constraints
Initialize population $Pop_{ij} \leftarrow rand_{ij}[L, H]$ and Evaluate fitness $Fit_j \leftarrow f(Pop_j)$
for $g = 1$ to GEN **do**
    for $j = 1$ to NP **do**
        Choose randomly $r_{1,2,3} \in [1, \ldots, NP]$, $r_1 \neq r_2 \neq r_3 \neq j$
        Create trial individual $X \leftarrow \mathcal{S}(r, F, Cr, Pop)$
        Verify boundary constraints **if** $(x_i \notin [L, H])$ $x_i \leftarrow rand_i[L, H]$
        Select better solution (X or Pop_j), and update $iBest$ if required
    end for
end for

1.5 Secrets of a Great Success

Now that you know the basic algorithm of differential evolution very well I would like to ask you a little question: How do we create the trial individual? You will certainly give me the right answer.[1] And profound comprehension of the answer to this question uncovers the secrets of differential evolution.

> **The success of differential evolution resides in the manner of the trial individual creation.**

Moreover, I boldly assure you that neither boundary constraint verification nor selection of better individuals are determining factors for DE. They, their varieties, and their properties are exceptionally good studied in other evolutionary methods such as genetic algorithms and evolution strategies. We can only adopt some elements according to our needs.

Here, in this section, I want to reveal a secret. And I shall do it with the aid of a graphical interpretation of the algorithm. First, I propose that you investigate the following figure (Fig. 1.6).

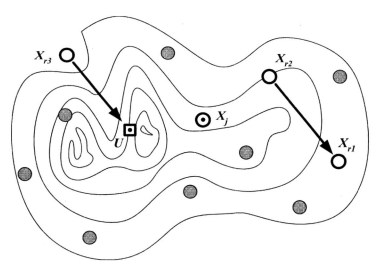

Fig. 1.6. Creation of the differential mutation vector U: three individuals are randomly chosen from the population; the scaled difference between two individuals is added to the third individual.

Suppose that there is some function of two variables outlined by its level lines. The population is initialized by a set of randomly generated individuals (circles on the figure). The algorithm is iterating and we are at some

---

[1] Prompting; see (1.2).

generation, where a certain jth individual X_j is under consideration. Three mutually different individuals, X_{r_1}, X_{r_2}, and X_{r_3}, have already been randomly chosen from the population. And we are naturally creating the intermediate differential mutation vector U:

$$U = X_{r_3} + F \cdot (X_{r_1} - X_{r_2}) . \tag{1.3}$$

Let us examine carefully the formula (1.3). The scaled difference between two randomly chosen individuals, $F \cdot (X_{r_1} - X_{r_2})$, defines direction and length of the search step. The constant F is a control parameter, which manages the trade-off between exploitation and exploration of the space. Then, this difference is added to the third randomly chosen individual, that serves as a base point of application (current reference point).

The fundamental idea of this formula and of differential evolution on the whole is to adapt the step length intrinsically along the evolutionary process.[2] At the beginning of generations the step length is large, because individuals are far away each from other. As the evolution goes on, the population converges and the step length becomes smaller and smaller. The randomness of both search directions and base points provides in many cases the global optimum, slightly retarding convergence.

> **The concept of differential evolution is a spontaneous self-adaptability to the function.**

Second, let me pass to Fig. 1.7. The creation of the trial individual X is outlined there. The individual under consideration (current individual) X_j underlies the trial one. Each parameter of the differential mutation vector U is accepted by the trial individual with some probability Cr. In order to insure that at least one parameter will be changed the random number Rnd generated in the range from 1 to D is used. Thus, the Rndth parameter of the differential mutation U is smoothly inherited by trial X. In Fig. 1.7 we arranged that $Rnd = 1$; it means that u_1 is unambiguously imitated by X: $x_1 \leftarrow u_1$, and the second and the last parameter, as we have a two-variable function, will be accepted with the probability Cr. Let us assume that in this case, $x_2 \leftarrow u_2$, and hence $X = U$.

Certainly we could use U directly for the next stage. But often the realization of this operation, creation of X, has a favorable effect: increasing of exploration/exploitation capabilities (diversity control) and handling of some functions' properties, that, in turn, results in better convergence of the algorithm.

---

[2] Such self-adaptability to the function's surface is clearly reflected in Covariance Matrix Adaptation Evolutionary Strategies (CMA-ES is one of the best state-of-the-art algorithms). And what is more, the DE method easily "learns" the surface of a function without any computational effort, whereas covariance matrix calculations, in CMA-ES, require perceptible exertion, in particular with increasing of the problem dimension.

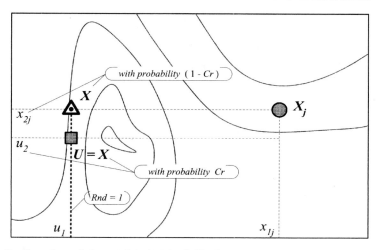

Fig. 1.7. Creation of the trial individual X: the current individual X_j is the basis of the trial one; each parameter of the differential mutation vector U is accepted with some probability Cr; at least one randomly chosen parameter Rnd is changed.

Third, it remains only to select the best individual. This operation is illustrated in Fig. 1.8. The trial individual is compared with the current one. If the trial has an equal or better fitness value, then it replaces the current one in the population. Thus, the population fitness is either always improving or at least keeping the same values. The described selection is called an elitist selection and is used successfully in many evolutionary algorithms.

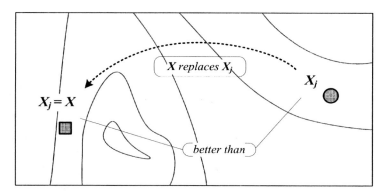

Fig. 1.8. Elitist selection: if the trial X is better than the current individual X_j, then it replaces the current one; the fitness of the population is always improving.

Summing up all the aforesaid, I emphasize the next three keys to DE success:

1. *Spontaneous self-adaptability*
2. *Diversity control*
3. *Continuous improvement*

Problems

1.1. Explain evolution in the full sense of the word? What do we mean when we speak about evolution from the algorithmic point of view?

1.2. What is an individual in the wide and narrow sense? How do we construct the individual in differential evolution?

1.3. How do we decide that one individual is "better" than another?

1.4. An engineering plant produces coil compression springs. Because of imported competing products the management of the plant decided to reduce the production cost by minimizing volume (weight) of the steel wires necessary for manufacturing a spring. A coil spring is described by three designing *variables*: (1) N – number of spring coils (takes only integer values), (2) D – outside diameter (any values), (3) d – wire diameter (strictly standardized, accepts values from a finite set). The spring must satisfy the following *constraints*: (a) the wire diameter is no smaller than d_{min}, (b) the outside diameter of the spring should be, at least, three times bigger than the wire diameter, to avoid the wound coils, (c) the outside diameter of the spring is no greater than D_{max}, (d) the free length l_f is no greater than l_{max}, (e) the allowable shear stress, $\frac{8C_f F_{max} D}{\pi d^3}$, is no greater than S, (f) the allowable deflection under preload $\sigma_p = F_p/K$ is no greater than σ_{max}; where $K = \frac{G \cdot d^4}{8 \cdot N \cdot D^3}$, $l_f = F_{max}/K + 1.1d(N + 2)$ and $C_f = \frac{4(D/d)-1}{4(D/d)-4} + \frac{0.6d}{D}$. F_{max} – maximum working load, S – allowable maximum shear stress, G – shear material module, F_p – preliminary compression force, l_{max} – maximum free length, d_{min} – minimum wire diameter, D_{max} – maximum outside diameter, σ_{max} – allowable maximum deflection under preload are engineering constants specifying the spring.

What is a criterion for optimization in this problem? Upon which parameters (variables) does it depend? Derive a formula for the chosen criterion for optimization. Create a prototype of the individual X for this problem. Outline the feasible region Ω of individuals, that is, the region where the variables do not violate the constraints. Take two feasible individuals $X_1, X_2 \in \Omega$ and compare them according to the derived criterion of optimization. You can assign any reasonable values for the engineering constants.

1.5. What is "population"? Roughly speaking, how big a population do you need to solve problem (1.4).

1.6. What is a mathematical model of a phenomenon? Propose two different mathematical models for problem (1.4). Which of the two models is better? Why?

1.7. What is optimization in the strict (engineering) sense of the word? Why do we need to optimize a model/problem? (From this point onwards I shall not differentiate when I speak about a model or a problem. It is implied that each problem has its model.) Formulate an optimization task for the preferred model found in (1.6). How many solutions do you think problem (1.4) has? How many if the outside diameter would be standardized? What is the global solution (optimum)? In what cases is the global optimum preferable? Is problem (1.4) the case?

1.8. What evolutionary algorithms do you know? What is artificial evolution?

1.9. Give a definition of the term "linear operator".

1.10. In what year did differential evolution appear for the first time?

1.11. What key element (procedure) distinguishes differential evolution from other population-based techniques?

1.12. What is a search strategy? Why do we need several different search strategies? What kinds of search strategies do you know?

1.13. Except differential mutation (or more general, differentiation) what evolutionary operations do you know?

1.14. As you can see, problem (1.4) may be formulated more naturally using simultaneously continuous, integer and discrete variables. We call it a mixed variable model. Is it possible to adapt the differential evolution algorithm to solve the problem with mixed variables?

1.15. How many constraints are in problem (1.4)? How many nonlinear constraints? Can you represent some of those constraints as boundary constraints? How many boundary constraints did you find?

1.16. Enumerate the species of differential evolution. In what situations does one prefer to use the transversal species?

1.17. Why does differential evolution require control parameters? How many control parameters has DE? List them.

1.18. What is hybridization and what are its uses? Describe each of four levels of hybridization. Give an example for each level.

1.19. How many global optima has a miltimodal function? Give an example of such a function. In what domains could multimodal optimization be necessary?

1.20. What is multiobjective optimization? Give an example of a problem where the multiobjective optimization is required.

1.21. Problem (1.4) is a problem of difficult optimization. Give a definition of the class of problems of difficult optimization. Which class of optimization methods is most suitable for solving these problems?

1.22. How do we define metaheuristics? List at least three advantages of meta-heuristic methods.

1.23. Are there some disadvantages of metaheuristics? If yes, enumerate them and show in which situations such a disadvantage causes essential difficulties of applying the method.

1.24. What is the difference between Simulated Annealing and Genetic Algorithms?

1.25. What population-based metaheuristics do you know? Point out, at least, three of their advantages as compared with neighbourhood metaheuristics.

1.26. Summarize for what optimization problems and under what conditions differential evolution is the best method of choice.

1.27. Recall and describe a real optimization problem you confronted in your life where you could efficiently apply differential evolution.

1.28. Enumerate at least eight application domains where differential evolution can be successfully implemented.

1.29. Program your own (quasi) random number generator. The simplest is a linear random number generator $x_k = a \cdot x_{k-1} + b$, where a and b are constants influencing the quality of a random sequence x_k. Take them as $a = 1664525$ and $b = 1013904223$ according to Knuth (1981). If you want to have the values lying in $(0, 1]$, you should divide x_k on 2^{32} for "unsigned long".

1.30. Estimate the quality of your random number generator (1.29). For this create a program where you generate 4000 random numbers between 0 and 1. Keep track of how many numbers are generated in each of the four quartiles, 0.0–0.25, 0.25–0.5, 0.5–0.75, 0.75–1.0. Compare the actual counts with the expected number. Is the difference within reasonable limits? How can you quantify whether the difference is reasonable?

1.31. Implement the famous differential evolution algorithm, if you have not yet done so. Use Rosenbrock's function for testing. Assign values to the control parameters NP, F and Cr. Run the optimization and obtain the optimal solution. Repeat this 10 times for different values of control parameters. Determine the best values for control parameters.

1.32. Write a short paragraph with sketches explaining the effect of self-adaptability to a function. Discuss how a chaotic behaviour becomes an intelligent one.

1.33. Test your differential evolution using Rastrigin's function (see Appendix C), the first time without crossover operation $Cr = 1$ and the second time with small values of crossover constant. Compare the results and discuss utility of the crossover operation.

1.34. Differential evolution uses a hard selection scheme (elitist selection). Implement the soft selection, where the new individual passes into the next generation with some probability (similarly to simulated annealing). In which cases would a soft selection be preferable? Give a real example.

2

Neoteric Differential Evolution

In this chapter you will make the acquaintance of the newest statement of differential evolution. But first, I propose that you dip into the background of population-based methods, so-called evolutionary algorithms. Also, I shall show the basic evolutionary algorithm scheme and apply it to differential evolution. After a rigorous mathematical definition of the optimization problem, I shall show you the fresh wording of the differential evolution algorithm, then together we shall compare it with the classical one, described in Chapter 1, and emphasize its advantages. In conclusion I shall point out the efficient techniques to handle mixed variables as well as constraints.

2.1 Evolutionary Algorithms

Let us talk a little more about Evolutionary Algorithms (EA). As you have already heard, these are population-based metaheuristics that can be applied with success in cases where traditional methods do not give satisfactory results. Originally inspired by the theory of evolution proposed by Darwin, these methods gave birth to the whole discipline, Evolutionary computation [SDB+93], that involves the simulation of natural evolution processes on a computer.

Evolutionary algorithms emerged in the sixties. Initially, EA were presented by three general trends. These are the genetic algorithms, evolution strategy, and evolutionary programming. Later, in the early nineties, the fourth trend, genetic programming, has come to light.

Genetic Algorithms. This is one of the most popular ideas in evolutionary computation. The concept of genetic algorithms was introduced and developed by J. Holland [Hol75]. In order to achieve a better understanding of the biological adaptation mechanisms he tried to simulate these processes numerically. That, in turn, resulted in the first genetic algorithm. Soon afterwards,

K. DeJong [DeJ75] formalized genetic algorithms for the binary search space. And some years later, thanks to D. Goldberg [Gol89], genetic algorithms became widely available.

Evolution Strategies were proposed by I. Rechenberg [Rec73] and H. Schwefel [Sch81]. Solving aviation engineering problems, for which classical optimization suffers a defeat, they revealed the most important key positions in evolutionary algorithms, namely, the ideas of adaptation and self-adaptation for control parameters of an algorithm.

Evolutionary Programming was elaborated by L.J. Fogel [FOW66]. Working on the evolution of finite state machines to predict time series, he gave birth to a new evolutionary branch. Being the result of, or to be more exact, the desire to procreate machine intelligence, evolutionary programming finally became an efficient optimizer. Later, this trend was appreciably enlarged by D.B. Fogel [Fog92].

Genetic Programming, successfully introduced by J. Koza [Koz92], arose from the evolution of more complex structures such as a set of expressions of a programming language and neural networks. J. Koza presented the structure (individual) in the form of trees, orientable graphs without cycles, in which each of the nodes is associated with a unit operation related to the problem domain.

For a deeper examination of this topic I definitely suggest a quite recent reference book of A.E. Eiben and J.E. Smith [ES03]. And now we shall consider the basic scheme that generalizes all evolutionary algorithms (see Fig. 2.1).

Evolutionary Algorithms. The vocabulary of evolutionary algorithms is, to a great extent, borrowed from both the biology and the theory of evolution. A set of problem parameters, *genes*, is described by an *individual*. An ensemble of individuals composes a *population*. Before optimizing, EA is initialized, often randomly because usually we do not have ideas about optimum localization, by a population of individuals. We name this *initialization*. Next, the optimization criterion, in the EA case so-called *fitness*, is calculated for each individual of the population. This is *evaluation*. Sometimes evaluation of fitness can be a computationally intensive operation. Thus, the initial population of *Parents* is ready and the algorithm begins its *evolutionary cycle*. Iterations, in EA terms, *generations*, last until a *stopping condition* is attained. In each evolutionary cycle the population passes through the following three steps.

1. *Selection* of the individuals that are more apt to reproduce themselves, from the population.
2. *Variations* of the selected individuals in a random manner. Mainly two operations are distinguished here: *crossover* and *mutation*. The variations of Parents germinate *Children*.
3. *Replacement* refreshes the population of the next generation usually by the best individuals chosen among Parents and Children.

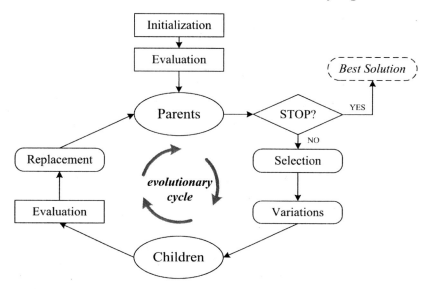

Fig. 2.1. A basic scheme of evolutionary algorithms.

Evolutionary Algorithm typically can be outlined by the following way (see Alg. 2).

Algorithm 2 Typical Evolutionary Algorithm

---

generation $g \leftarrow 0$
population $\mathbb{P}^g \leftarrow$ *Initialize*
fitness $f(\mathbb{P}^g) \leftarrow$ *Evaluate*
while (**not** stopping condition) **do**
   // *proceed to the next evolutionary cycle* //
   $g \leftarrow g + 1$
   Parents $\leftarrow$ *Select* from $\mathbb{P}^g$
   Children $\leftarrow$ *Vary* Parents (*Crossover, Mutation, ...*)
      fitness $\leftarrow$ *Evaluate* Children
   Replacement $\mathbb{P}^g \leftarrow$ *Survive* Parents and Children
end while

2.2 Problem Definition

The overall goal of an optimization problem $f : M \subseteq \mathbb{R}^D \to \mathbb{R}, M \neq \emptyset$, where f is called the *objective function* (also *fitness* or *cost function*), is to find a vector $X^* \in M$ such that:

$$\forall X \in M : f(X) \geq f(X^*) = f^*, \tag{2.1}$$

where f^* is called a *global minimum*; X^* is the *minimum location* (*point* or *set*).

$$M = \{X \in \mathbb{R}^D \mid g_k(X) \leq 0, \forall k \in \{1, \dots, m\}\} \tag{2.2}$$

is the set of feasible points for a problem with inequality constraints $g_k : \mathbb{R}^D \to \mathbb{R}$.

A particular case of inequality constraints is boundary constraints

$$L \leq X \leq H : \quad L, H \in \mathbb{R}^D. \tag{2.3}$$

For an unconstrained problem $M = \mathbb{R}^D$.

Because $\max\{f(X)\} = -\min\{-f(X)\}$, the restriction to minimization is without loss of generality. In general the optimization task is complicated by the existence of nonlinear objective functions with multiple local optima. A *local minimum* $\hat{f} = f(\hat{X})$ is defined by the condition (2.4).

$$\exists \epsilon : \forall X \in M \mid \|X - \hat{X}\| < \epsilon \Rightarrow \hat{f} \leq f(X). \tag{2.4}$$

2.3 Neoteric Differential Evolution

As with all evolutionary algorithms, differential evolution deals with a population of solutions. The population $\mathbb{P}$ of a generation g has NP vectors, so-called individuals of population. Each such individual represents a potential optimal solution.[1]

$$\mathbb{P}^g = \{X_i^g\}, \qquad i = 1, \dots, NP. \tag{2.5}$$

In turn, the individual contains D variables, so-called *genes*.

$$X_i^g = \{x_{i,j}^g\}, \qquad j = 1, \dots, D. \tag{2.6}$$

Usually, the population is initialized by randomly generating individuals within the boundary constraints (2.3),

$$\mathbb{P}^0 = \{x_{i,j}^0\} = \{rand_{i,j} \cdot (h_j - l_j) + l_j\}, \tag{2.7}$$

where the *rand* function uniformly generates values in the interval $[0, 1]$.

---

[1] In order to show the flexibility of implementation, here I represent a population and an individual as a set of elements instead of a vector presentation.

Then, for each generation all the individuals of the population are updated by means of a reproduction scheme. Thereto for each individual ind a set π of other individuals is randomly extracted from the population. To produce a new one the operations of differentiation and crossover are applied one after another. Next, selection is used to choose the best. Let us consider these operations in detail.

First, a set of randomly extracted individuals $\pi = \{\xi_1, \xi_2, \ldots, \xi_n\}$ is necessary for differentiation. The strategies (i.e., a difference vector δ and a base vector β) are designed on the basis of these individuals. Thus, the result of differentiation, the so-called *trial* individual, is

$$\omega = \beta + F \cdot \delta\,, \tag{2.8}$$

where F is the constant of differentiation. I shall show an example of a typical strategy [SP95]. Three different individuals are randomly extracted from the population. The trial individual is equal to $\omega = \xi_3 + F \cdot (\xi_2 - \xi_1)$ with the difference vector $\delta = \xi_2 - \xi_1$ and the base vector $\beta = \xi_3$.

Afterwards, the trial individual ω is recombined with the target one ind. Crossover represents a typical case of a gene's exchange. A new trial individual inherits genes of the target one with some probability. Thus,

$$\omega_j = \begin{cases} \omega_j & \text{if } rand_j \geq Cr \\ ind_j & \text{otherwise} \end{cases} \tag{2.9}$$

where $j = 1, \ldots, D$, $rand_j \in [0, 1)$ and $Cr \in [0, 1]$ is the constant of crossover. This was a combinatorial crossover. Also, other types of crossover can be used: binary approach [SP95], mean-centric (UNDX, SPX, BLX), and parent-centric (SBX, PCX) approaches [DJA01].

Selection is realized by simply comparing the objective function values of target and trial individuals. If the trial individual better minimizes the objective function, then it replaces the target one. This is the case of *elitist* or so-called "greedy" selection.

$$ind = \begin{cases} \omega & \text{if } f(\omega) \leq f(ind) \\ ind & \text{otherwise}\,. \end{cases} \tag{2.10}$$

Notice that there are only three control parameters in this algorithm. These are:

- NP – population size
- F – constant of differentiation
- Cr – constant of crossover

As for stopping conditions, one can either fix the number of generations $g_{\max}$ or a desirable precision of the solution VTR (*value-to-reach*).

The pattern of the DE algorithm is presented hereafter (see Alg. 3).

Algorithm 3 Neoteric Differential Evolution

Require: F, Cr, NP – control parameters
   initialize $\mathbb{P}^0 \leftarrow \{ind_1, \ldots, ind_{NP}\}$
   evaluate $f(\mathbb{P}^0) \leftarrow \{f(ind_1), \ldots, f(ind_{NP})\}$
   while (**not** stopping condition) **do**
      for all $ind \in \mathbb{P}^g$ **do**
         $\mathbb{P}^g \rightarrow \pi = \{\xi_1, \xi_2, \ldots, \xi_n\}$
         $\omega \leftarrow Differentiation(\pi, F, Strategy)$
         $\omega \leftarrow Crossover(\omega, Cr)$
         $ind \leftarrow Selection(\omega, ind)$
      end for
      $g \leftarrow g + 1$
   end while

2.4 Distinctions and Advantages

Above all, I would like to compare differential evolution to the basic EA scheme. As you have already observed, initialization and evaluation are kept without changes. A general EA provides for Darwin's mechanism of parent selection, however, where more apt it stands a better chance to reproduce itself; differential evolution applies variations (differentiation and crossover) sequentially to each individual. For that, an ensemble of individuals is randomly chosen from the population each time. The result of variations is child, called a trial individual. Moreover, in DE the trial immediately replaces its ancestor in the population if its fitness is better than or equal to its ancestor's. Also, a stopping condition is verified right here after replacement. Finally, for more clarification I propose that you familiarize yourself with the individual's cycle in differential evolution, which I presented in Fig.2.2.

Furthermore, let us touch nicety and compare neoteric differential evolution with the classical one, described in Chapter 1.

The first point that is significant is the dissociation of differentiation and crossover in the new DE statement. Of course, we can always associate back these two operations as soon as we need them. However, such a dissociation offers us self-evident advantages. This is, at the minimum, the independent study and use of the operations that enables us to exclude one operation from another and thoroughly analyze their behavior and influence on the search process. Next, differential mutation of classic DE is generalized to differentiation of a neoteric one. From a theoretical point of view, it gives the unlimited spectrum of strategies that obey the unique and universal principle of optimization $\omega = \beta + F \cdot \delta$. In practice, we can now manipulate with great ease the search strategies according to the needs of an optimization task.

The second significant point is the crossover operation in itself. Now the basic element of crossover is a trial individual created in the issue of differentiation, rather than a current one. This improvement changes the philosophy

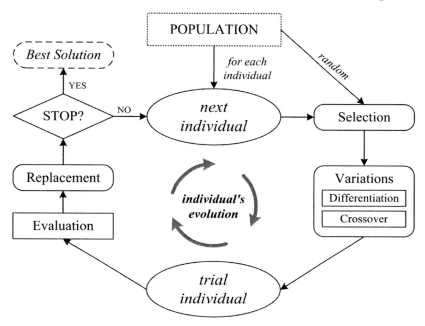

Fig. 2.2. The evolutionary cycle of an individual in differential evolution.

of the solution search. If before the variation operations are considered as the mutation of a current individual resembling evolutionary strategies, then now the main attention is completely focused on the creation of a new, more perfect individual. Such an individual is produced primarily on the basis of the actual state of the population and certainly may inherit some properties of a current individual.

Also, I moved away the mutation parameter $Rnd \in [1, \dots, D]$ (see (1.2)). I do not consider it very important for optimization. Besides, we can almost always imitate it by the appropriate choice of the crossover value Cr. For example, following (2.9), $Cr = 0$ (absence of crossover) $\Rightarrow$ the new-created individual is completely inherited; $Cr = 1 \Rightarrow$ the current individual is completely inherited; and $Cr \approx 1 - 1/D$ permits us to inherit the minimal number of the new-created individual's genes. Although it does not guarantee absolutely that at least one new gene passes into the next generation (the case of classical DE), it certainly guarantees that there is a great chance it does happen.

Here I emphasized three **principal advantages** of the new algorithm's statement. Perhaps, in future, continuously working with differential evolution you will find many more advantages then I did. So, these are as follows.

1. *Efficiency.* A special stress is laid on the efficient creation of a new member of a population, instead of the mutation of current individuals.

2. *Flexibility.* The new algorithm is more flexible to use and adapts to modification; it is preferred for research purposes. In particular, the isolation in the reproduction cycle of differentiation, crossover, and selection from one another in action allows natural and easy driving by the evolution process.
3. *Fundamentality.* The well-known algorithm stated in Chapter 1 is just a particular case of neoteric differential evolution. In fixing the differentiation strategy and appropriate crossover, variation operations can be convoluted to a single equation similar to (1.2). Moreover, differentiation synthesizes in itself the fundamental ideas in optimization.[2] The operation of Differentiation intrinsically generalizes the universal concepts of the solution search, as in the case of traditional versus modern methods of optimization.

2.5 Mixed Variables

Differential evolution in its initial form is a method for continuous variable optimization [SP95]. However, in [LZ99b, LZ99c] the DE modification for integer and discrete variables is proposed. We first show **integer variable** handling.

Despite the fact that DE works with continuous values on the bottom level, for the evaluation of the objective function integer values are used. Thus,

$$
\begin{aligned}
f = f(Y): \quad & Y = \{y_i\} \\
\text{where} \quad & y_i = \begin{cases} x_i & \text{for continuous variables} \\ \lfloor x_i \rfloor & \text{for integer variables} \end{cases}, \\
\text{and} \quad & X = \{x_i\}, \quad i = 1, \dots, D.
\end{aligned}
\tag{2.11}
$$

The $\lfloor x \rfloor$ function gives the nearest integer less than or equal to x. Such an approach provides a great variety of individuals and ensures algorithm robustness. In other words, there is no influence from discrete variables on algorithm functioning. In the integer case, the population initialization occurs as follows.

$$
\mathbb{P}^0 = \{x_{i,j}^0\} = \{rand_{i,j} \cdot (h_j - l_j + 1) + l_j\}, \qquad rand_{i,j} \in [0, 1). \tag{2.12}
$$

Next, **discrete variables** can be designed in the same easy way. It is supposed that a discrete variable $Z^{(d)}$ takes its values from the discrete set $\{z_i^{(d)}\}$ containing l ordered elements.

$$
\begin{aligned}
Z^{(d)} = \{z_i^{(d)}\}, \qquad & i = 1, \dots, l \\
\text{so that} \quad z_i^{(d)} < & z_{i+1}^{(d)}.
\end{aligned}
\tag{2.13}
$$

---

[2] I mean here an iterative procedure of choosing a new base point, direction, and optimization step. I shall illustrate it in detail in the next chapter (Chapter 3).

Instead of the discrete values $z_i^{(d)}$ their indexes i are used. Now, the discrete variable $Z^{(d)}$ can be handled as an integer one with boundary constraints $(1 \leq i \leq l)$. For the evaluation of the objective function the discrete value itself is used in place of its index. Thus, discrete variable optimization is reduced to the integer variable one and discrete values are used only for objective function evaluation.

This approach showed the best results among evolutionary algorithms for mixed variable problems in mechanical engineering design [Lam99].

2.6 Constraints

Let the optimization problem be presented in the generally used form (2.14).

$$
\text{find} \quad X^* : f(X^*) = \min_X f(X)
$$

subject to

boundary constraints $\quad L \leq X \leq H$ (2.14)

constraint functions $\quad g_k(X) \leq 0, \qquad k = 1, \ldots, m$

2.6.1 Boundary Constraints

Boundary constraints represent low and high limits put on each individual:

$$
L \leq \omega \leq H . \tag{2.15}
$$

It is necessary that new values of variables satisfy the constraints after differentiation (or reproduction). For that, the values that have broken range conditions are randomly put back inside their limits.

$$
\omega_j = \begin{cases} rand_j \cdot (h_j - l_j) + l_j & \text{if} \quad \omega_j \notin [l_j, h_j] \\ \omega_j & \text{otherwise} \end{cases} \tag{2.16}
$$

$$
j = 1, \ldots, D .
$$

For integer variables one uses the next modification of this equation:

$$
\omega_j = \begin{cases} rand_j \cdot (h_j - l_j + 1) + l_j & \text{if} \quad \lfloor \omega_j \rfloor \notin [l_j, h_j] \\ \omega_j & \text{otherwise} \end{cases} . \tag{2.17}
$$

In addition to the reinitialization (2.16) there are also other ways of boundary constraint handling. For example:

- Repeating of differentiation (2.8) until the trial individual satisfies the boundary constraints,

- Or, use of the periodic mode or the shifting mechanism proposed in [ZX03, MCTM04, PSL05],
- Or else, taking into consideration that boundary constraints are inequality constraints ($L - X \leq 0$ and $X - H \leq 0$), constraint handling methods and their modifications developed for constraint functions are also suitable for handling boundary constraints.

2.6.2 Constraint Functions

Penalty Function Method

I shall represent here a penalty function method for constraint handling used by Lampinen and Zelinka [LZ99b, LZ99c]. Compared with hard constraint handling methods, where infeasible solutions are rejected, the penalty function method uses penalties for moving into the feasible area M (2.2). Such penalties are directly added into the objective function. We give it in a logarithmic form:

$$\log \tilde{f}(\omega) = \log \left(f(\omega) + a \right) + \sum_{k=1}^{m} b_k \cdot \log c_k(\omega)$$

$$c_k(\omega) = \begin{cases} 1.0 + s_k \cdot g_k(\omega) & \text{if} \quad g_k(\omega) > 0 \\ 1.0 & \text{otherwise} \end{cases} \qquad (2.18)$$

$$s_k \geq 1.0$$
$$b_k \geq 1.0$$
$$\min f(\omega) + a > 0 \,.$$

It is necessary that the objective function take only nonnegative values. For this reason the constant a is added. Even if the constant a takes too high values it does not affect the search process. The constant s scales constraint function values. The constant b modifies the shape of the optimizing surface. When the function value for a variable that lies outside a feasible area is insignificant, it is necessary to increase the values s and b. Usually, satisfactory results are achieved with $s = 1$ and $b = 1$.

It is clear that this method demands the introduction of extra control parameters, and therefore, in order to choose their effective values additional efforts are necessary. Generally, it is realized by trial and error, when the algorithm is started repeatedly for many times under various parameter values (s, b). It is obvious that this is not effective enough, so researchers are continuing investigations in this domain.

Modification of the Selection Operation

An original approach for constraint problem solution has been proposed in [Lam01, MCTM04]. The selection rule modification (2.10), where there is no need of using the penalty functions, has been shown there.

The basic idea is applying multiobjective optimization for handling constraints. This idea, it seems, was first communicated by David Goldberg as early as 1992 [Deb01] (pp.131–132). Later, three of its instances sequentially were reported to wider audience: (1) Coello Coello [Coe99b], (2) Deb [Deb00], and (3) Lampinen [Lam01]. Below, I shall describe Lampinen's instance [Lam01], which is based on pure pareto-dominance defined in constraint function space.[3]

The choice of individual results from the next three rules.

- If both solutions ω and ind are feasible, preference is given to the lower objective function solution.
- The feasible solution is better than the infeasible.
- If the both solutions are infeasible, preference is given to the less infeasible solution.

Mathematically these rules are written as:

$$ind = \begin{cases} \omega & \text{if} \quad \Phi \vee \Psi \\ ind & \text{otherwise} \end{cases},$$

where

$$\Phi = [\forall k \in \{1, \ldots, m\} : g_k(\omega) \leq 0 \wedge g_k(ind) \leq 0] \wedge \qquad (2.19)$$
$$\wedge [f(\omega) \leq f(ind)]$$
$$\Psi = [\exists k \in \{1, \ldots, m\} : g_k(\omega) > 0] \wedge$$
$$\wedge [\forall k \in \{1, \ldots, m\} : \max(g_k(\omega), 0) \leq \max(g_k(ind), 0)] .$$

Thus, the trial vector ω will be chosen if:

- It satisfies all the constraints and provides lower objective function value or
- It provides lower than or equal to ind value for all constraint functions.

Notice that in the case of an infeasible solution, the objective function is not evaluated.

To prevent stagnation [LZ00], when the objective function values of both trial and target vectors are identical, preference is given to the trial one. In Appendix B you can find (proposed by me) the C source code of the above-described selection rules.

Other Constraint-Handling Methods

Finally I shall present a general classification of the constraint-handling methods for evolutionary algorithms. More detailed information can be found in [MS96, Coe99a, Coe02].

---

[3] Discussed in personal communication with J. Lampinen.

1. Methods based on preserving feasibility of solutions
   - Use of specialized operators (Michalewicz and Janikow, 1991)
   - Searching the boundary of feasible region (Glover, 1977)
2. Methods based on penalty functions
   - Method of static penalties (Homaifar, Lai and Qi, 1994)
   - Method of dynamic penalties (Joines and Houck, 1994)
   - Method of annealing penalties (Michalewicz and Attia, 1994)
   - Method of adaptive penalties (Bean and Hadj-Alouane, 1992)
   - Death penalty method (Bäck, 1991)
   - Segregated genetic algorithm (Le Riche, 1995)
3. Methods based on a search for feasible solutions
   - Behavioral memory method (Schoenauer and Xanthakis, 1993)
   - Method of superiority of feasible points (Powell and Skolnick, 1993)
   - Repairing infeasible individuals (Michalewicz and Nazhiyath, 1995)

Problems

2.1. What does evolutionary computation study?

2.2. What three general trends of development of evolutionary algorithms do you know?

2.3. Review once again problem (1.4). Indicate the "genes" of the individual.

2.4. Usually, in differential evolution, the population is initialized by random values within boundary constraints. Propose your technique of initialization for the following cases: (a) there are no boundary constraints; (b) you have one or two solutions, but you do not know if these solutions are optimal; (c) your constraint handling method requires only feasible individuals and you need to preserve the uniformity of initialization. Implement and test the proposed techniques.

2.5. Evaluate the fitness of the function $f(X) = e^{x_1 x_2 x_3 x_4 x_5} - \frac{1}{2}(x_1^3 + x_2^3 + 1)^2$ for the individual $X_0 = (-1.8, 1.7, 1.9, -0.8, -0.8)$.

2.6. Which of the Evolutionary Algorithm's operations is/are, most probably, time-consuming? (a) selection, (b) crossover, (c) mutation, (d) differentiation, (e) variation, (f) evaluation, (g) replacement, (h) recombination, (i) initialization. Explain your answer.

2.7. What elements are common for all evolutionary algorithms?

2.8. What is the difference between iteration, generation and evolutionary cycle?

2.9. What is the stopping condition? Propose three different ways to end the optimization. Add it in your differential evolution and test.

2.10. You are given the following maximization problem.

$$\max \quad \frac{a x_1^4 x_3^2}{\pi^2 x_2^3 x_4} - \cos^2(2\pi d \frac{x_5}{x_3}) + e^{b \sin(2x_1)/x_2^3} - 3\ln(c\frac{\pi}{4}x_2^2) + x_1 x_5$$

$$x_1 + x_2 + x_3 + x_4 \leq x_5$$

$$x_1, x_2, x_3, x_4, x_5 \geq 0$$

$$a \geq 0, \; b < 0, \; c > 0, \; d \leq 0$$

Transform this problem into a minimization problem.

2.11. How many optima has the function $|sin(x)|$ in the range from 0 to 10?

2.12. Give a definition of local optimum.

2.13. Does the function given in problem (2.11) have a global optimum?

2.14. How does one calculate the trial individual? Show the concrete formula and give an explanatory sketch.

2.15. Determine and explain what is a "parent" and what is a "child" in differential evolution?

2.16. The crossover operation executes the inheritance of genes from the old to the new individual. Develop and implement your own crossover instead of the formula (2.9) proposed in Chapter 2.

2.17. What is the minimal size of the population you can use according to the formula you demonstrated in problem (2.14)?

2.18. Does differential evolution obey the natural selection theory of Darwin? What are the common and distinguishing features?

2.19. Is it possible, in differential evolution, that the "child" becomes "parent" in one and the same generation?

2.20. Find four distinctions between the classical DE (famous algorithm) and the neoteric one.

2.21. Recall and explain three principal advantages of neoteric DE.

2.22. Explain how DE handles integer variables? What is the advantage as against gradient methods?

2.23. Solve by hand the Traveling Salesman Problem with Time Windows. A truck driver must deliver to 9 customers on a given day, starting and finishing in the depot. Each customer $i = 1, \ldots, 9$ has a time window $[b_i, e_i]$ and an unloading time u_i. The driver must start unloading at client i during the specified time interval. If he is early, he has to wait till time b_i before starting to unload. Node 0 denotes the depot, and c_{ij} the time to travel between nodes i and j for $i, j \in \{0, 1, \ldots, 9\}$. The data are $u = (0, 1, 5, 9, 2, 7, 5, 1, 5, 3)$, $b = (0, 2, 9, 4, 12, 0, 23, 9, 15, 10)$, $e = (150, 45, 42, 40, 150, 48, 96, 100, 127, 66)$, and

$$
(c_{ij}) = \begin{pmatrix}
- & 5 & 4 & 4 & 4 & 6 & 3 & 2 & 1 & 8 \\
7 & - & 2 & 5 & 3 & 5 & 4 & 4 & 4 & 9 \\
3 & 4 & - & 1 & 1 & 12 & 4 & 3 & 11 & 6 \\
2 & 2 & 3 & - & 2 & 23 & 2 & 9 & 11 & 4 \\
6 & 4 & 7 & 2 & - & 9 & 8 & 3 & 2 & 1 \\
1 & 4 & 6 & 7 & 3 & - & 8 & 5 & 7 & 4 \\
12 & 32 & 5 & 12 & 18 & 5 & - & 7 & 9 & 6 \\
9 & 11 & 4 & 12 & 32 & 5 & 12 & - & 5 & 22 \\
6 & 4 & 7 & 3 & 5 & 8 & 6 & 9 & - & 5 \\
4 & 6 & 4 & 7 & 3 & 5 & 8 & 6 & 9 & -
\end{pmatrix}
$$

2.24. Find an efficient model of problem (2.23) for solving by differential evolution. Focus attention on data representation, especially on realization of the permutation of clients and the fitness function. Solve the problem by DE and compare the results.

2.25. Solve problem (1.4) supposing that d and D take only integer values.

2.26. Write a code to handle discrete variables. Apply it to solving the problem (1.4) as is.

2.27. Formulate the problem of placing N queens on an N by N chessboard such that no two queens share any row, column, or diagonal. Use binary variables.

2.28. Could DE optimize binary variables? If yes, write the proper code and solve the N-queens problem (2.27) for $N = 4, 8, 16, \ldots$. Otherwise, use a permutation to model the problem and solve it in integer variables. Think about an efficient method of constraints handling. Compare the results and determine which of two methods is more clever. Explain why.

2.29. What are the boundary constraints? What methods to handle boundary constraints do you know? Point out at least four methods and explain by giving an example.

2.30. Elaborate your own method of boundary constraints handling. Estimate its influence on the convergence of algorithms using the test functions from Appendix C.

2.31. The solution to a system of nonlinear equations specified by a mapping $f : \mathbb{R}^n \to \mathbb{R}^n$ is a vector $X \in \mathbb{R}^n$ such that $f(X) = 0$. Algorithms for systems of nonlinear equations usually approach this problem by seeking a local minimizer to the problem

$$\min \{\|f(X)\| \; : \; L \leq X \leq H\},$$

where $\|\cdot\|$ is some norm on $\mathbb{R}^n$, most often the l_2 norm. Solve any reasonable system of nonlinear equations using your own method of handling boundary constraints.

2.32. What is the penalty function? Create the penalty function for problem (1.4).

2.33. Solve problem (2.32). Experiment on the parameters s_k and b_k of the penalty function. Try to find their optimal values. Estimate its influence on the algorithm performance.

2.34. What drawbacks do you see in using penalty methods?

2.35. Try to implement independently a modification of the selection operation. Solve problem (1.4) with this method. Compare the results with ones obtained in problem (2.33).

2.36. Given D electrons, find the equilibrium state distribution of the electrons positioned on a conducting sphere. This problem, known as the Thomson problem of finding the lowest energy configuration of D point charges on a conducting sphere, originated with Thompson's plum pudding model of the atomic nucleus. The potential energy for D points (x_i, y_i, z_i) is defined by

$$f(x, y, z) = \sum_{i=1}^{D-1} \sum_{j=i+1}^{D} \left((x_i - x_j)^2 + (y_i - y_j)^2 + (z_i - z_j)^2 \right)^{-\frac{1}{2}},$$

and the constraints on the D points are

$$x_i^2 + y_i^2 + z_i^2 = 1, \quad i = 1, \dots, D.$$

The number of local minima increases exponentially with D. Theoretical results show that

$$\min\{f(p_1, \dots, p_D) : \|p_i\| = 1, 1 \le i \le D\} \ge \frac{1}{2} D^2 (1-\epsilon), \quad 0 \le \epsilon \le \left(\frac{1}{D} \right)^{1/2}.$$

Solve this problem for $D = 3, 10, 50$. How do you handle the nonlinear equality constraints? Are you far from the theoretical results?

2.37. Choose from the list of constraint-handling methods at the end of Subsection 2.6.2 any method you please and implement it.

3

Strategies of Search

In this chapter[1] we shall discuss the most important operation of differential evolution — *differentiation*. The differentiation operation can be realized by many search strategies. At the beginning I shall illustrate the strategies (schemes) that were proposed before my research work. Then, starting from common principles for all methods of continuous optimization, I introduce the unique and universal formula of differentiation. The given formula classifies all the strategies of DE into the four groups according to their search behavior: random, directed, local, and hybrid. Here, in this chapter, you will find the strict description of all groups of strategies. And for each group I shall demonstrate a set of practical examples. All these examples cover, from a practical point of view, almost all possible variants of strategies. Finally, I shall speak about new functions of differentiation constant F and test results.

3.1 Antecedent Strategies

There are five DE strategies (or schemes) that were proposed by K. Price and R. Storn [Sto96a]:

- Scheme DE/rand/1 $\omega = x_1 + F \cdot (x_2 - x_3)$
- Scheme DE/rand/2 $\omega = x_5 + F \cdot (x_1 + x_2 - x_3 - x_4)$
- Scheme DE/best/1 $\omega = x_{best} + F \cdot (x_1 - x_2)$
- Scheme DE/best/2 $\omega = x_{best} + F \cdot (x_1 + x_2 - x_3 - x_4)$
- Scheme DE/rand-to best/1 $\omega = x_{ind} + \lambda \cdot (x_{best} - x_1) + F \cdot (x_2 - x_3)$

Later, two more strategies were introduced by Fan and Lampinen:

---

[1] Parts of this chapter are based on material that originally appeared in [FJ04g, FJ04d].

- Trigonometric scheme [FL01]
  $\omega = (x_1 + x_2 + x_3)/3 + (p_2 - p_1) \cdot (x_1 - x_2) + (p_3 - p_2) \cdot (x_2 - x_3) + (p_1 - p_3) \cdot (x_3 - x_1)$
  $p_i = |f(x_i)/(f(x_1) + f(x_2) + f(x_3))|, \quad i = 1, 2, 3;$

- Directed scheme [FL03]
  $\omega = x_3 + (1 - f(x_3))/f(x_1) \cdot (x_3 - x_1) + (1 - f(x_3))/f(x_2) \cdot (x_3 - x_2),$
  where $f(x_3) \leq f(x_1), \quad f(x_3) \leq f(x_2).$

3.2 Four Groups of Strategies

What is the common point for all the methods of continuous optimization? Perhaps that we take some initial point, *base* point, and from this point we search some *direction* in which we suppose attaining the optimum as soon as possible. Personally, I began from this assumption. And I desired to simulate this principle with differential evolution. For that I searched a certain operation in the form of $\omega = \beta + F \cdot \delta$, where β, *base* vector, and δ, *difference* vector, are calculated with consideration of the actual population state (see Fig. 3.1). Afterward, I baptized this operation *differentiation*.

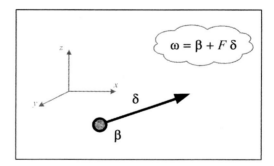

Fig. 3.1. Fundamental principle inherent to all methods of continuous optimization applied to differential evolution: β – base point, δ – optimal direction.

But how we can create this β and δ? The most practical solution is to use the barycenters of the individuals randomly chosen from the population. I accepted two variants to create β: random and local. In the latter case I gave preference to the best individual of the population. As for δ, either one constructs two barycenters in a random manner, or constructs them taking into consideration the values of an objective function, so the difference vector is oriented or directed. I would like to note that the directed case interprets well the simulation of the gradient. All these combinations of β and δ choice lead us to the four groups of strategies: random, directed, local, and hybrid (see Fig. 3.2).

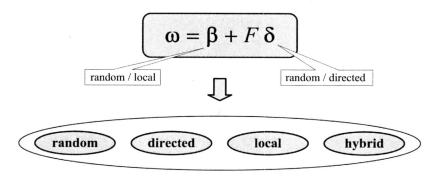

Fig. 3.2. Four groups of strategies.

So, by using the knowledge about values of the objective function I have divided all the possible strategies into four groups:

1. *RAND group* contains the strategies in which the trial individual is generated without any information about values of the objective function.
2. *RAND/DIR group* consists of the strategies that use values of the objective function to determine a "good" direction. This group represents an imitation of the gradient function.
3. *RAND/BEST group* uses the best individual to form the trial one. It looks like a chaotical local search around the current best solution.
4. *RAND/BEST/DIR group* combines the last two groups into one including all their advantages.

Notice that the introduction of information such as the "good" direction and the best individual considerably reduces the search space exploration capabilities (see Chapter 4). Thus, it makes the algorithm more stable, but on the other hand, sometimes the algorithm becomes less efficient for complex problems, for example, with many local optima. So, the practical question of what strategy to choose is always defined by the concrete problem and the concrete objective function. This fact is theoretically proven in the *No Free Lunch* theorem [WM95, WM97]:

> *All algorithms that search for an extremum of an objective function perform exactly the same, according to any performance measure, when averaged over all possible objective functions. In particular, if algorithm A outperforms algorithm B on some objective functions, then loosely speaking there must exist exactly as many other functions where B outperforms A.*

Let us now examine these groups of strategies.

3.2.1 RAND Group

Randomly extracted individuals X_i are arbitrarily separated into two classes C' and C'' containing n' and n'' elements accordingly. Then the barycenters of these classes $V_{C'}$ and $V_{C''}$ are found by the formula

$$V_C = \frac{1}{n} \sum_{i=1}^{n} X_i \qquad n = n', n''. \tag{3.1}$$

There are two possibilities for choosing the base vector $\beta = V_g$:

1. Using some individual from these classes $V_g \in C' \cup C''$;
2. Using another individual from the population $V_g \notin C' \cup C''$.

Thus, the differentiation formula for this group of strategies is (Fig. 3.3)

$$\omega = V_g + F \cdot (V_{C''} - V_{C'}) \tag{3.2}$$

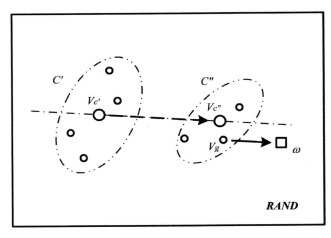

Fig. 3.3. *RAND* group of strategies.

3.2.2 RAND/DIR Group

Let randomly extracted individuals X_i be divided into two classes C_+ and C_- with n_+ and n_- elements so, that for each element from the class C_+ its objective function value would be less than the objective function value of any element from class C_-. That is,

$$f(X_i) \leq f(X_j) \quad : \quad (\forall X_i \in C_+) \wedge (\forall X_j \in C_-)$$
$$i = 1, \ldots, n_+, \qquad j = 1, \ldots, n_- \, . \tag{3.3}$$

We find then the maximal and minimal elements of each of the classes

$$f(X_{C_+}^{\min}) \leq f(X_i) \leq f(X_{C_+}^{\max}), \qquad \forall X_i \in C_+$$
$$f(X_{C_-}^{\min}) \leq f(X_i) \leq f(X_{C_-}^{\max}), \qquad \forall X_i \in C_- \, . \tag{3.4}$$

Then we calculate the positive and negative shifts inside the classes.

$$V_s^{\pm} = \lambda \cdot \left(X_{C_\pm}^{\min} - X_{C_\pm}^{\max} \right)$$
$$\lambda = 1/2 \quad - \text{ influence constant} \, . \tag{3.5}$$

So the average shift is equal to

$$V_S = (V_s^+ + V_s^-)/2 \, . \tag{3.6}$$

Hence, the differentiation formula is

$$\omega = V_{C_+} + F \cdot (V_{C_+} - V_{C_-} + V_S) \, , \tag{3.7}$$

where V_{C_+} and V_{C_-} are barycenters of C_+ and C_- accordingly (see Fig. 3.4).

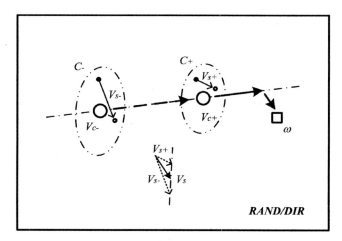

Fig. 3.4. *RAND/DIR* group of strategies.

3.2.3 RAND/BEST Group

Randomly extracted individuals X_i such as in the *RAND* group are divided into two classes C' and C'' with n' and n'' elements correspondingly. But as the base vector β the current best individual V_b is used here. So, the differentiation formula is (see Fig. 3.5)

$$\omega = V_b + F \cdot (V_{C''} - V_{C'}).$$ (3.8)

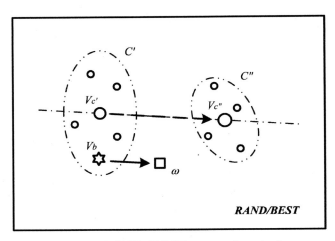

Fig. 3.5. *RAND/BEST* group of strategies.

3.2.4 RAND/BEST/DIR Group

In addition to the direction information the best individual is taken into account. The division of extracted individuals into two groups and the finding of their barycenters and the average shift are the same as in the *RAND/DIR* case. Thus, the differentiation formula is (see Fig. 3.6)

$$\omega = V_b + F \cdot (V_{C_+} - V_{C_-} + V_S).$$ (3.9)

3.2.5 On the Constant of Differentiation

The constant of differentiation F is one of the control parameters that considerably affects the convergence rate. In the first works on DE [Sto96a, LZ99b] it was recommended to vary F in the $(0, 2+]$ range. Because of the generalization of strategies proposed here I changed this range to the new limits:

$$F \in (-1, 0) \cup (0, 1+].$$ (3.10)

In this way differentiation can point in both directions. The negative interval, $F \in (-1, 0)$, restricts us to local search between the barycenters. The positive interval, $F \in (0, 1+]$, extends the exploration of the search space.

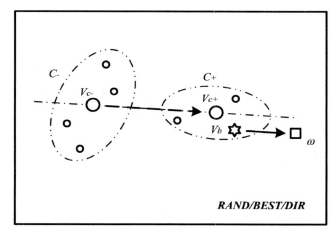

Fig. 3.6. _RAND/BEST/DIR_ group of strategies.

3.3 Examples of Strategies

In this section we shall consider detailed examples of strategies that form each of the groups. We shall set in the particular Differentiation formulae for these strategies and deduce the interrelation between the general differentiation constant F and its particular representation φ according to the strategy.

At first, I shall fix the notation:

- ind — target (current) individual.
- $\{x_i\}$ — set of extracted individuals.
- $V^* \in \{x_i\}$ is an individual that has the minimal value of the objective function among extracted individuals and the target one.
- V' and V'' are other individuals; if there are more than three extracted individuals we denote them $V_1, V_2, V_3, \ldots$.
- V_b — the best individual.
- δ — difference vector.
- φ — particular constant of differentiation.
- β — base vector, the point of the difference vector application.

So, the formula of differentiation always has the following form;

$$\omega = \beta + \varphi \cdot \delta \qquad (3.11)$$

3.3.1 RAND Strategies

Rand1 Strategy

In this strategy only one random individual x_1 is extracted from the population. At the same time this individual presents the base vector β of the strategy. The difference vector δ is formed by the current and extracted individuals $(x_1 - ind)$. The step length is equal to $\| \varphi \cdot (x_1 - ind) \|$. The formula of differentiation for this strategy is

$$w = x_1 + \varphi \cdot (x_1 - ind) . \tag{3.12}$$

Comparing with the main group's formula (3.2) we can see that $V_g = V_{C''} = x_1$ and $V_{C'} = ind$. Therefore, the constant of differentiation $F = \varphi$. Such an approach gives $(NP-1)$ possibilities for the trial individual, where NP is the size of population. This strategy is shown in Fig. 3.7.

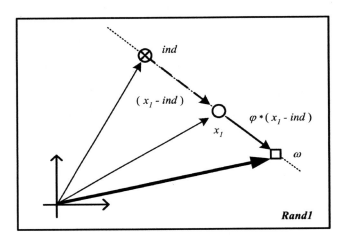

Fig. 3.7. *Rand1* strategy.

Rand2 Strategy

The target individual *ind* interchanges with a second randomly extracted one x_2. Thus, the number of possible trial individuals increases to $(NP-1)(NP-2)$. As we can see, by extracting an additional individual the exploration capabilities augment considerably. The difference is $\delta = x_1 - x_2$, and the base is $\beta = x_1$, where x_1 and x_2 represent the barycenters $V_{C'}$ and $V_{C''}$ accordingly. $V_g = x_1$ and $F = \varphi$. The differentiation formula of this strategy is

$$w = x_1 + \varphi \cdot (x_1 - x_2) . \tag{3.13}$$

The strategy is shown in Fig. 3.8.

Rand3 Strategy

Here, three individuals are randomly extracted to form this strategy. So, all possible combinations are augmented to $(NP-1)(NP-2)(NP-3)$. The first

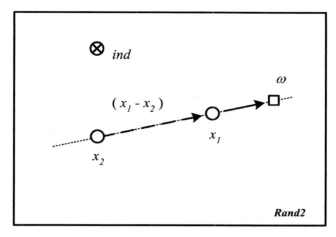

Fig. 3.8. *Rand2* strategy.

two extracted vectors generate the difference $\delta = x_1 - x_2$, but in this case the base vector is the third extracted vector x_3. The formula of differentiation is

$$\omega = x_3 + \varphi \cdot (x_1 - x_2) . \tag{3.14}$$

By assumption that $V_g = x_3$, $V_{C'} = x_2$, and $V_{C''} = x_1$ the strategy can be generalized to the common group representation (3.2). Also, $F = \varphi$. The strategy is shown in Fig. 3.9.

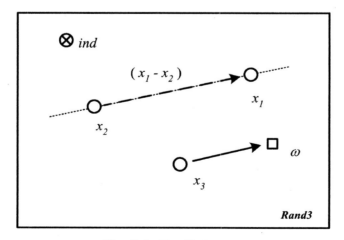

Fig. 3.9. *Rand3* strategy.

Rand4 Strategy

By extracting four random individuals the next strategy can be written as

$$\omega = x_2 + \varphi \cdot (x_2 - x_1 + x_4 - x_3) , \tag{3.15}$$

where the difference $\delta = x_2 - x_1 + x_4 - x_3$ and the base $\beta = x_2$. In this case, the difference is generated by superposition of two random directions $(x_2 - x_1)$ and $(x_4 - x_3)$. Any of these individuals can be chosen as the base vector: for example, $\beta = x_2$ as in (3.15). The vectors x_1 and x_3 create the barycenter $V_{C'} = (x_1 + x_3)/2$ and the same with the vectors x_2 and x_4: $V_{C''} = (x_2 + x_4)/2$. $V_g = x_2$. So, (3.15) may be rewritten as $\omega = x_2 + \varphi \cdot (x_2 - x_1 + x_4 - x_3) = x_2 + 2\varphi \cdot ((x_2 + x_4)/2 - (x_1 + x_3)/2) = V_g + 2\varphi \cdot (V_{C''} - V_{C'})$. It is clear that $F = 2\varphi \Rightarrow \varphi = F/2$. We can see that in order to harmonize the strategy with its template it is necessary to divide F by 2. The number of possible combinations of four random individuals is $(NP - 1)(NP - 2)(NP - 3)(NP - 4)$. This strategy is shown in Fig. 3.10.

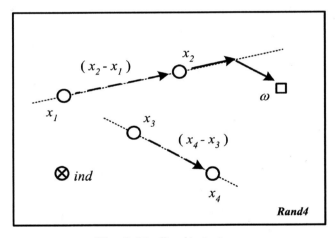

Fig. 3.10. *Rand4* strategy.

Rand5 Strategy

In this strategy five random individuals are extracted. The first four individuals work as in the *Rand4* strategy, and the fifth individual realizes the base point $\beta = V_g = x_5$. Notice that $F = \varphi/2$ too. The search space exploration becomes appreciable; there are $(NP - 1)(NP - 2)(NP - 3)(NP - 4)(NP - 5) = \prod_{i=1}^{5}(NP - i)$ potential combinations of extracted vectors. The strategy is shown in Fig. 3.11.

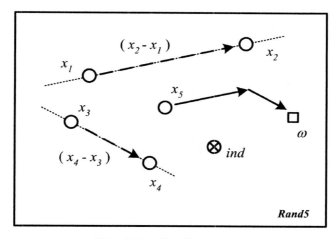

Fig. 3.11. *Rand5* strategy.

3.3.2 RAND/DIR Strategies

The main principle remains the same as in the RAND group of strategies. Moreover, the information about the objective function is used to calculate the direction of differentiation. In this way, the probability of the optimal choice increases twice. Such an approach is analogous to descent in a direction opposite to the gradient vector.

Rand1/Dir1 Strategy

This strategy uses one extracted individual x_1 and target individual ind for the differentiation formula. With introduced notations it looks like

$$\omega = V^* + \varphi \cdot (V^* - V') , \tag{3.16}$$

or, equivalently

$$\omega = \begin{cases} x_1 + \varphi \cdot (x_1 - ind) & \text{if} \quad f(x_1) < f(ind) \\ ind + \varphi \cdot (ind - x_1) & \text{otherwise} . \end{cases}$$

This formula is similar to its generalization with $V_{C_+} = V^*$ and $V_{C_-} = V'$. Hence $F = \varphi$ and $V_S = 0$. The strategy is shown in Fig. 3.12.

Rand2/Dir1 Strategy

Following such a tendency I shall illustrate the strategy in which two random individuals x_1 and x_2 are extracted. The formula of this strategy is the same

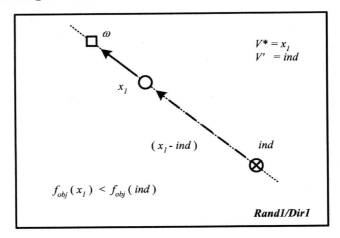

Fig. 3.12. *Rand1/Dir1* strategy.

as (3.16), but the current individual *ind* is not used. It may be represented also as

$$\omega = \begin{cases} x_1 + \varphi \cdot (x_1 - x_2) & \text{if} \quad f(x_1) < f(x_2) \\ x_2 + \varphi \cdot (x_2 - x_1) & \text{otherwise} \end{cases} . \tag{3.17}$$

Also, $F = \varphi$ and $V_S = 0$. The strategy is shown in Fig. 3.13.

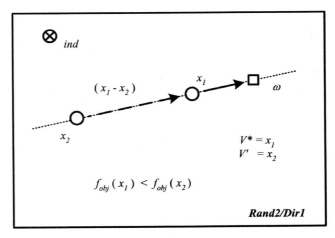

Fig. 3.13. *Rand2/Dir1* strategy.

Rand3/Dir2 Strategy

The next strategy uses three random elements x_1, x_2, and x_3. These elements are sorted according to their values of the objective function so that $f(V^*) \leq f(V')$ and $f(V^*) \leq f(V'')$, where $V^*, V', V'' \subseteq \{x_1, x_2, x_3\}$. The formula of this strategy is:

$$\omega = V^* + \varphi \cdot (2V^* - V' - V'') . \tag{3.18}$$

Note that the superposition of two random directions $(V^* - V')$ and $(V^* - V'')$ (difference vector) is used here. The base point β is the individual with the minimal value of the objective function V^*. To adjust this differentiation's formula with its template (3.7) imagine that the individuals form two barycenters $V_{C_-} = (V' + V'')/2$ and $V_{C_+} = V^*$. Thus, $\omega = V^* + 2\varphi \cdot (V^* - (V' + V'')/2) = V_{C_+} + 2\varphi \cdot (V_{C_+} - V_{C_-})$. There is no average shift vector in this case ($V_S = 0$). So, $\varphi = F/2$. The strategy is shown in Fig. 3.14.

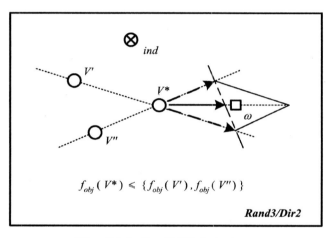

Fig. 3.14. *Rand3/Dir2* strategy.

Rand3/Dir3 Strategy

This strategy completely repeats the previous one. In addition to it, the average shift vector V_S is formed by comparing the objective function values of two other individuals V' and V''. Assume that $f(V^*) \leq f(V') \leq f(V'')$; then the differentiation formula is

$$\omega = V^* + \varphi \cdot \left(2V^* - \frac{1}{2}V' - \frac{3}{2}V''\right) . \tag{3.19}$$

Supposing $V_{C_-} = (V' + V'')/2$ and $V_{C_+} = V^*$; the shift vectors for each of the barycenters are equal to $V_s^+ = 0$ and $V_s^- = (V' - V'')/2$. Then the average shift is $V_S = (V_s^+ + V_s^-)/2 = (V' - V'')/4$. By substituting it in the template (3.7) we obtain $\omega = V_{C_+} + F \cdot (V_{C_+} - V_{C_-} + V_S) = V^* + F \cdot (V^* - (V' + V'')/2 + (V' - V'')/4) = V^* + F/2 \cdot (2V^* - (1/2)V' - (3/2)V'')$. Thus, $\varphi = F/2$. This strategy is shown in Fig. 3.15.

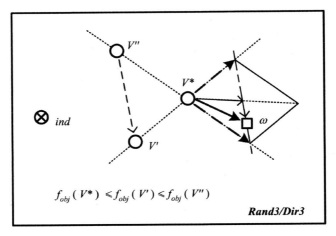

Fig. 3.15. *Rand3/Dir3* strategy.

Rand4/Dir2 Strategy

This strategy is based on the choice of four random individuals x_1, x_2, x_3, and x_4. Let us denote these individuals as V_1, V_2, V_3, and V_4 so that $f(V_1) \leq f(V_2)$ and $f(V_3) \leq f(V_4)$. Hence the differentiation formula is

$$\omega = V_1 + \varphi \cdot (V_1 - V_2 + V_3 - V_4). \tag{3.20}$$

It is obvious that $V_{C_+} = (V_1 + V_3)/2$ and $V_{C_-} = (V_2 + V_4)/2$. The only distinction is that the base vector $\beta = V_1$, but not V_{C_+} as in the template (3.7). Such a small difference allows us to simplify the strategy without losing quality. Moreover it is easy to verify that $\varphi = F/2$ and $V_S = 0$. This strategy is shown in Fig. 3.16.

Rand4/Dir3 Strategy

This strategy continues to evolve the ideas of the *Rand3/Dir2* one. Here, it is applied in three directions constructed on four randomly extracted individuals. V^*, the individual with the minimal value of the objective function, presents the positive barycenter V_{C_+}. The other three individuals form the negative one $V_{C_-} = (V_1 + V_2 + V_3)/3$. Thus, the formula is:

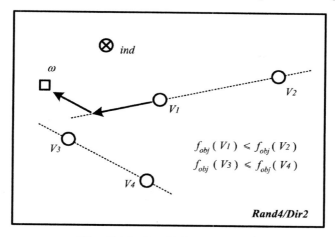

Fig. 3.16. *Rand4/Dir2* strategy.

$$\omega = V^* + \varphi \cdot (3V^* - V_1 - V_2 - V_3).$$ (3.21)

It is obvious that $\varphi = F/3$ and $V_S = 0$. This strategy is shown in Fig. 3.17.

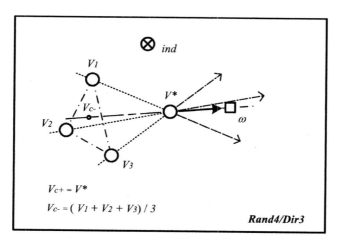

Fig. 3.17. *Rand4/Dir3* strategy.

Rand5/Dir4 Strategy

Continuing to follow the tendency, we illustrate the strategy built on five random individuals, which form four random directions. The purpose of increasing the number of extracted individuals is to determine more precisely the descent direction by better exploiting the information about the objective function and, at the same time, to explore the search space more and more entirely. The differentiation formula, that represents this case is

$$\omega = V^* + \varphi \cdot (3V^* - V_1 - V_2 - V_3 - V_4) \,. \qquad (3.22)$$

Here, $\varphi = F/4$, $V_S = 0$. The strategy is shown in Fig. 3.18.

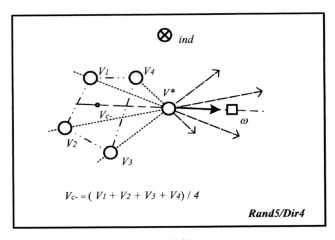

Fig. 3.18. *Rand5/Dir4* strategy.

3.3.3 RAND/BEST Strategies

This group of strategies is characterized by a random local search around the best individual of the population. The main principle is similar to the *RAND* group, but the base vector β, in this case, is always the best individual V_b. Such an approach appropriately provides a local search, especially when the gradient of the objective function tends to zero, and the gradient methods suffer a defeat.

Rand1/Best Strategy

In this strategy one random individual x_1 is extracted. The difference vector δ is formed by the current *ind* and extracted x_1 individuals. Then this difference is added to the best individual V_b. The formula of differentiation is

$$\omega = V_b + \varphi \cdot (x_1 - ind) \,; \qquad (3.23)$$

$\varphi = F$. The strategy is shown in Fig. 3.19.

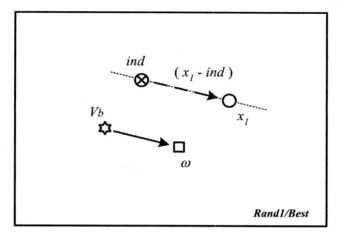

Fig. 3.19. *Rand1/Best* strategy.

Rand2/Best Strategy

Two randomly extracted individuals are used here. The current one does not participate in forming the difference vector. The differentiation formula is:

$$\omega = V_b + \varphi \cdot (x_2 - x_1).\qquad(3.24)$$

Hence, $\varphi = F$. The strategy is shown in Fig. 3.20.

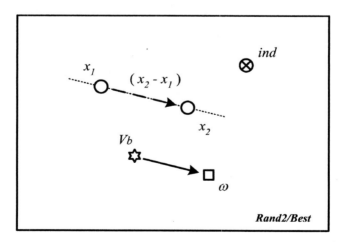

Fig. 3.20. *Rand2/Best* strategy.

Rand3/Best Strategy

Three random individuals with the current one create two random directions. Each pair makes barycenters $V_{C'} = (ind + x_2)/2$ and $V_{C''} = (x_1 + x_3)/2$ accordingly. The differentiation formula for this strategy is

$$\omega = V_b + \varphi \cdot (x_1 - ind + x_3 - x_2). \tag{3.25}$$

By comparing with template (3.8) it is clear that $\varphi = F/2$. The strategy is shown in Fig. 3.21.

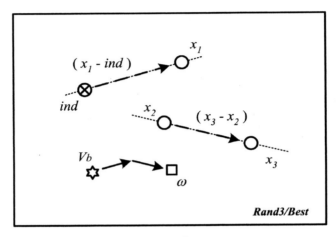

Fig. 3.21. *Rand3/Best* strategy.

Rand4/Best Strategy

The next strategy also creates two random directions but without the current individual. Four random elements are extracted from the population for this purpose. The rest completely coincide with the previous strategy *Rand3/Best*. The formula of differentiation is:

$$\omega = V_b + \varphi \cdot (x_1 - x_2 + x_3 - x_4). \tag{3.26}$$

Also, $\varphi = F/2$. The strategy is shown in Fig. 3.22.

3.3.4 RAND/BEST/DIR Strategies

The combination of two previous groups *RAND/DIR* and *RAND/BEST* generates this group of strategies. The differentiation in the descent direction and the simultaneous local search around the best solution are incorporated together.

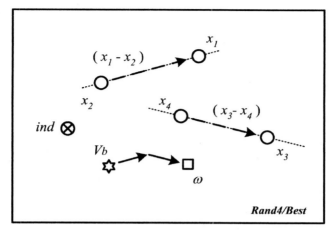

Fig. 3.22. *Rand4/Best* strategy.

Rand0/Best/Dir1 Strategy

In this strategy there are no random individuals. The trial individual is created by the both target and best individuals. Such maximal simplification reduces computing time. The differentiation formula is:

$$\omega = V_b + \varphi \cdot (V_b - ind) . \tag{3.27}$$

$\varphi = F$. The strategy is shown in Fig. 3.23.

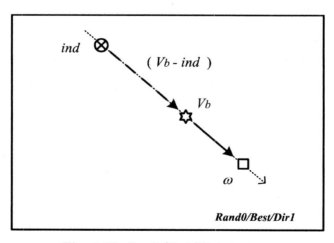

Fig. 3.23. *Rand0/Best/Dir1* strategy.

Rand1/Best/Dir1 Strategy

In this strategy the best individual V_b with the extracted one x_1 form the positive barycenter V_{C_+}. The negative barycenter is presented by the current individual ind. In other words, the descent direction from the current ind to the best V_b individual is perturbed by randomly extracted x_1. The differentiation formula is:

$$\omega = V_b + \varphi \cdot (V_b + x_1 - 2ind) . \tag{3.28}$$

So, $\omega = V_b + \varphi \cdot (V_b + x_1 - 2ind) = V_b + 2\varphi \cdot ((V_b + x_1)/2 - ind) = V_b + 2\varphi \cdot (V_{C_+} - V_{C_-})$. Thereby, $\varphi = F/2$. The strategy is shown in Fig. 3.24.

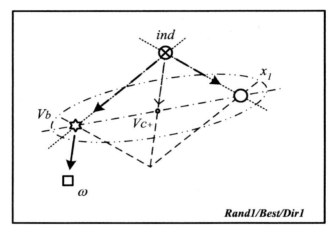

Fig. 3.24. *Rand1/Best/Dir1* strategy.

Rand1/Best/Dir2 Strategy

Here, the randomly extracted individual participates in the creation of the negative barycenter V_{C_-} as well as the current one ind. The positive barycenter V_{C_+} is presented by the best individual V_b. The formula of differentiation is

$$\omega = V_b + \varphi \cdot (2V_b - x_1 - ind) . \tag{3.29}$$

Likewise in the previous case, $\omega = V_b + \varphi \cdot (2V_b - x_1 - ind) = V_b + 2\varphi \cdot (V_b - (ind + x_1)/2) = V_b + 2\varphi \cdot (V_{C_+} - V_{C_-})$. So, $\varphi = F/2$. The strategy is shown in Fig. 3.25.

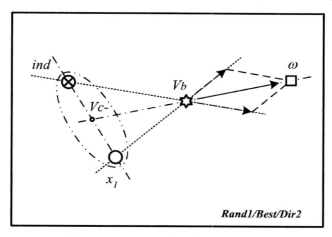

Fig. 3.25. *Rand1/Best/Dir2* strategy.

Rand2/Best/Dir1 Strategy

Two random individuals x_1 and x_2 are extracted in this strategy. The direction constructed on these vectors $(x_1 - x_2)$ randomly perturbs the main descent from the current individual ind to the best one V_b. The differentiation formula is

$$\omega = V_b + \varphi \cdot (V_b - ind + x_1 - x_2) \,. \tag{3.30}$$

It is supposed that $V_{C_-} = (ind + x_2)/2$ and $V_{C_+} = (V_b + x_1)/2$. Hence, $\omega = V_b + \varphi \cdot (V_b - ind + x_1 - x_2) = V_b + 2\varphi \cdot ((V_b + x_1)/2 - (ind + x_2)/2) = V_b + 2\varphi \cdot (V_{C_+} - V_{C_-})$. In this way, $\varphi = F/2$ too. The strategy is shown in Fig. 3.26.

Rand2/Best/Dir3 Strategy

Another way to use two random individuals is to create the negative barycenter V_{C_-} with the current one by randomly making three directions to the best individual V_b. The differentiation formula for this case is

$$\omega = V_b + \varphi \cdot (3V_b - ind - x_1 - x_2) \,. \tag{3.31}$$

It is easy to verify that $\varphi = F/3$. The strategy is shown in Fig. 3.27.

Rand3/Best/Dir4 Strategy

This strategy follows the same tendency, but here, three elements x_1, x_2, and x_3 are extracted to create the negative barycenter V_{C_-}. The current individual

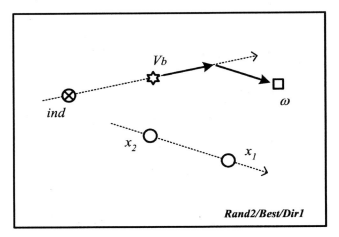

Fig. 3.26. *Rand2/Best/Dir1* strategy.

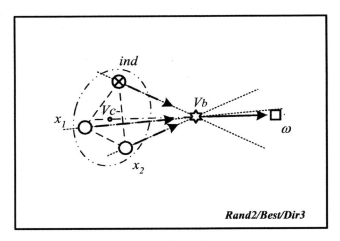

Fig. 3.27. *Rand2/Best/Dir3* strategy.

ind also participates in this strategy. So, there are four random directions that form the difference vector δ. The differentiation formula is

$$\omega = V_b + \varphi \cdot (4V_b - ind - x_1 - x_2 - x_3).\tag{3.32}$$

Let $V_{C_-} = (ind + x_1 + x_2 + x_3)/4$ and $V_{C_+} = V_b$, so $\omega = V_b + 4\varphi \cdot (V_{C_+} - V_{C_-})$. Hence, $\varphi = F/4$. The strategy is shown in Fig. 3.28.

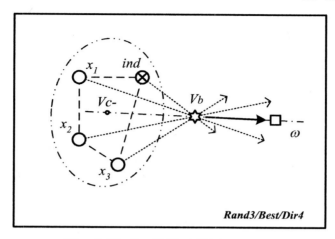

Fig. 3.28. *Rand3/Best/Dir4* strategy.

Rand4/Best/Dir4 Strategy

In this strategy there is no influence from the current individual *ind*. But the four directions are also used. For this purpose four random elements x_1, x_2, x_3, and x_4 are extracted. These vectors form the negative barycenter V_{C_-}. The differentiation formula is

$$\omega = V_b + \varphi \cdot (4V_b - x_1 - x_2 - x_3 - x_4) \, . \tag{3.33}$$

It is obvious that $\varphi = F/4$. The strategy is shown in Fig. 3.29.

Rand4/Best/Dir5 Strategy

To increase the search space exploration the current individual *ind* is added to the previous strategy *Rand4/Best/Dir4*. This strategy follows the tendency of the *Rand3/Best/Dir4* one, but the superposition of five directions is applied here in order to create the difference vector δ. The differentiation formula is

$$\omega = V_b + \varphi \cdot (5V_b - ind - x_1 - x_2 - x_3 - x_4) \, . \tag{3.34}$$

$\varphi = F/5$. The strategy is shown in Fig. 3.30.

3.4 Tests

All strategies presented here have been thoroughly tested on standard benchmarks (see Appendix C) with problem dimensions from 2 to 100 and a systematic range of control parameters. The obtained results are summarized.

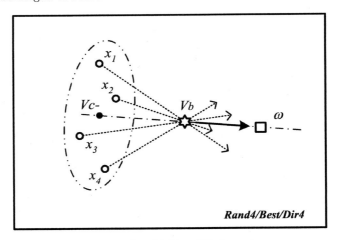

Fig. 3.29. *Rand4/Best/Dir4* strategy.

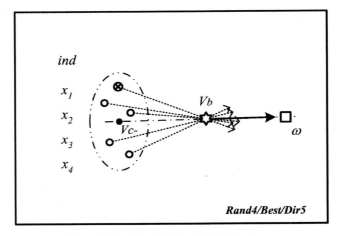

Fig. 3.30. *Rand4/Best/Dir5* strategy.

The full versions are kept in Excel and Text files (51.3 Mb), separately for each test function and for each dimension. Also, there is a short version. The short version contains the extraction of the best results and has been realized as a PDF document. All this information can be naturally obtained by simply contacting me. Taking into consideration the huge volumes of data, I did not dare place it in the book.

From these tests we can observe the following unquestionable facts:

1. There are strategies less sensitive to the control parameters. Generally, these are the strategies that are similar to arithmetic crossover. And there

are strategies more sensitive to the control parameters. These are basically the strategies that realize the local search around the best found solution. The practice shows that the more sensitive strategy is, the better the solution that can be achieved.

2. With increasing the dimensionality of a problem, increasing of number of randomly extracted individuals is needed. It could be explained by the need to increase of the dimensionality of subspaces, where the search is performed.

3. The more complex the function is, the more random strategy is needed.

Problems

3.1. Which operation is most important in differential evolution?

3.2. Explain search strategy? Why do we need to have so many strategies?

3.3. What is the trigonometric scheme? Write its formula and give a geometrical interpretation of how it works. Indicate advantages and disadvantages of this strategy. Implement it in your DE algorithm and compare with classical strategy $\omega = \xi_3 + F \cdot (\xi_2 - \xi_1)$ using the standard benchmarks (see Appendix C).

3.4. What is the common point for all methods of continuous optimization?

3.5. Give an explanation for β and δ. How can we create them taking into consideration the current state of the population? Propose your own method and implement it.

3.6. Which of the two, β or δ, could interpret the gradient's behaviour?

3.7. Classify all possible variants of behaviour of β and δ into four groups. Make a supposition as to which of the groups of strategies better corresponds to one or another test function.

3.8. How can one make the algorithm be more stable?

3.9. What is the meaning of the No Free Lunch theorem? And what do you think about the strategies now?

3.10. Given a $D \times D$ Hilbert matrix $\mathbf{H}$: $h_{ij} = 1/(i + j + 1)$, $i, j = 0, 1, \ldots, D - 1$. Find the inverse Hilbert matrix $\mathbf{H}^{-1}$. Because $\mathbf{H}$ is ill defined, it is very difficult to accurately compute $\mathbf{H}^{-1}$ as D increases. But the task can be formulated as the minimization problem

$$\min \sum_{i,j=0}^{D-1} |e_{ij}| , \quad \mathbf{E} = (e_{ij}) = \mathbf{I} - \mathbf{H}\mathbf{A} ,$$

where $\mathbf{I}$ is the identity matrix and $\mathbf{A}$ is an approximation of the inverse Hilbert matrix. You need to choose one strategy from each group of strategies that is most suitable, in your opinion, for solving this problem.

3.11. Code the minimization problem (3.10) and four chosen strategies. Solve the problem for $D = 3, 5, 10$, compare the results and discuss the strategies' behaviour.

$$\text{answer for} \quad D = 3 , \quad \mathbf{H}^{-1} \approx \mathbf{A}^* = \begin{pmatrix} 9 & -36 & 30 \\ -36 & 192 & -180 \\ 30 & -180 & 180 \end{pmatrix}$$

3.12. What value did you give to the constant of differentiation F in problem (3.11)? Try different values of F and observe the changes in results (optimum and convergence). How do the results change with increasing of F? For each strategy plot graphs of the algorithm's convergence for different F.

3.13. When should we take F in the interval $(-1, 0)$?

3.14. Which of your strategies, used in problem (3.12), is less sensible to the changes of F? And which is more sensible? Explain why?

3.15. How does the sensibility of a strategy influence the quality of solution in terms of convergence and precision?

4

Exploration and Exploitation

A successful application of an optimizer resides in the well-found trade-off between exploration and exploitation. So, we are continuously searching for the best equilibrium between them. In this chapter we pass to the analysis of the differentiation operation and equally to the study of the control parameter influence. In order to make a better choice of strategy I propose calculating an indicator of the strategy diversity,[1] its exploration capacity. Also, I shall show that differentiation is the first step to the general operator integrating mutation and crossover, where mutation provides the needed diversity of the population and crossover assures the capacities to survive. Moreover, in this chapter I expose my studies consecrated to the control parameters and their tuning. This results in practical recommendations for using differential evolution.

When we speak about evolutionary algorithms — GA, ES, DE, or others — we always expect to find the global optimum, but...

> ...the ability of an EA to find a global optimal solution depends on its ability to find a right relation between exploitation of the elements found so far and exploration of the search space.... [Bey98]

Thus, the successful application of the method consists in the choice of the optimal exploration/exploitation union. As is well known, the excessiveness of exploration leads to the global optimum with a high probability, but critically slows down the convergence rate. On the other hand, the excessive exploitation quickly results in local optima.

The capability of genetic operators to control exploration/exploitation balance as well as their relative importance has been discussed for many decades. Some groups of scientists believe in mutation-selection superiority, others concentrate themselves on crossover power. But I make an effort to be impartial to these opinions and elicit the advantages from both points of view.

---

[1] The diversity measures the proportion of the surveyed space of solutions.

4.1 Differentiation via Mutation

The strategies, which use objective function values to create the trial individual, accomplish an exploitation function. These are *dir* and *dir-best* groups. The diversity in this case decreases twice, so in order to maintain it at a required level it is necessary to increase the population size and/or the number of extracted individuals.

The fact of random choice of parents for a trial individual itself answers for exploration capabilities. Besides the population size and the type of strategy, exploration efficiency can be controlled by the differentiation constant F as well [Zah01, LL02b, Š02].

To the present day, it was considered (disruption and construction theories) that mutation cannot completely fulfill the functions of crossover and vice versa [Spe93]. Mutation perfectly creates a random diversity, but it cannot execute the construction function well. Crossover can show preservation, survival, and construction, but often it cannot achieve a desirable diversity. Thus, the EC community was looking forward to the one general operator that could integrate mutation and crossover functions as well as any variations between them. Differentiation in the DE algorithm is the first step on the road to such an operator. It does not fall under the influence of accepted disruption theory providing needed diversity and, at the same time, it luxuriously preserves survival capabilities. Let discuss it in more detail.

Differentiation is the first step to the general operator.

There is no disruption effect for differentiation! Disruption rate theory estimates the probability that an evolutionary operator will disrupt a hyperplane sample, in other words, the probability that individuals within a hyperplane will leave that hyperplane [Spe93]. Let all individuals of a population X_i belong to hyperplane H. Hence, β and δ are always on H. Therefore, $\omega = \beta + F \cdot \delta$ will belong H too. That is, there is no combination of individuals on the hyperplane that makes the trial individual leave this hyperplane. This means a good survival capability of differentiation usually inherent to crossover.

4.2 Crossover

The principal role of crossover is as a construction. There is no such mutation that can achieve higher levels of construction than crossover [Spe98]. Just as selection exploits objective function values, crossover exploits genetic information. Moreover, crossover furnishes the high diversity of a population.

Convinced of the power of crossover I would like to make a point about applying it to DE.

Videlicet, in the cases when we use the strategies with a direction analysis (*dir* and *dir-best* groups) crossover operation becomes unnecessary, because it spoils a trial individual inducing the noise. In other words, when we choose the directed strategies, it is supposed that we want to imitate the gradient function, that is, to make the steps close to the gradient direction. If we use crossover, the gene's exchange between the trial and target individuals would perturb the desired direction in most cases.

Furthermore, note that differentiation by itself is capable of executing the both functions (exploration/exploitation) simultaneously. So, if we guaranteed sufficient exploration (diversity of population), then the crossover operation would be superfluous. Thus we could eliminate it and thereby reduce computing time as well.

4.3 Analysis of Differentiation

The structure of the DE algorithm is similar to that of genetic algorithms: concepts of mutation and crossover are repeated here. In addition, DE integrates the ideas of self-adaptive mutation specific to evolution strategies. Namely, the manner of realization of such a self-adaptation has made DE one of the most popular methods in evolutionary computation. We examine it in detail.

Originally, two operations were distinguished: differential mutation and continuous recombination [Pri99]. Differential mutation was based on the strategy $\omega = \xi_3 + F \cdot (\xi_2 - \xi_1)$ and required at least three randomly extracted individuals. A continuous recombination was in need of only two individuals, $\omega = \xi_1 + K \cdot (\xi_2 - \xi_1)$. Price emphasized the different dynamic effects of these operations. In the case of a continuous recombination the trial individual ω places only on the line created by its parents ξ_1, ξ_2. This compresses a population. In the case of a differential mutation the difference vector $(\xi_2 - \xi_1)$ is applied to an independent individual. And it is similar to the Gaussian or Cauchy distribution used in ES, that makes no reference to the vector to which it is applied. It does not compress a population. Founded on such an inference several strategies were proposed [Sto96a].

Recently, in 2004, a new vision of these operations was discovered (see [FJ04d] or Chapter 3). A new principle of strategy design (see [FJ04g] or Section 3.2) was introduced, which synthesizes the previous two operations by one unique formula and accentuates population diversity. Now, all strategies are described by two vector terms: difference vector δ and base vector β (2.8). The difference vector provides a mutation rate term (i.e., a self-adaptive noise), which is added to a randomly selected base vector in order to produce a trial individual. The self-adaptation results from the individuals' positions. During the generations the individuals of a population occupy more and more profitable positions and regroup themselves. So, the difference vector decreases (automatically updates) each time the individuals fit local or global optima.

The strategies have been classified into four groups by information that they use to "differentiate" the actual individual (rand/dir/best/dir-best). Each group represents a proper method of search (random/directed/local/ hybrid). Hence, the key to ensure a required diversity of a population is not only the dynamic effect of the operations, but, to a greater extent, the number of randomly extracted individuals k needed to create a strategy.

We look at differentiation now from a combinatorial point of view. Usually, population size NP and the constant of differentiation F are fixed. Thus NP individuals are capable of producing potentially $\Theta(k)$ different solutions. We refer to $\Theta(k)$ as a diversity characteristic of a strategy, whereas the method of using of these individuals reflects strategy dynamics (see Sections 3.2 and 3.4 or [FJ04b]). We shall give an estimation of diversity. Let the individuals be extracted from the population one after another, so the upper diversity bound can be evaluated in the following way (see Fig. 4.1),

$$\overline{\Theta}(k) = \prod_{i=1}^{k}(NP - i). \tag{4.1}$$

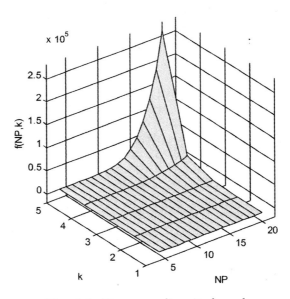

Fig. 4.1. The upper diversity bound.

The strategy (difference vector) is created by calculating the barycenters of two sets of individuals. The general differentiation formula can be rewritten as

$$\omega = \beta + F \cdot (Bar(set_2) - Bar(set_1)). \tag{4.2}$$

These sets, besides randomly extracted individuals (n_1 in the first set and n_2 in the second set), may also include the target and the current best individuals. Because of such a generalization, the extraction order of individuals forming a barycenter is not important, thus the diversity of population decreases in $n_1! \cdot n_2!$ times, where $n_1 + n_2 = k$. Moreover, if the directed or hybrid strategies are used, then the diversity still drops down twice. Therefore we introduce a direction factor

$$dir = \begin{cases} 2 & \text{if RAND/DIR or RAND/BEST/DIR} \\ 1 & \text{if RAND or RAND/BEST} \end{cases} \tag{4.3}$$

Consequently, the exact diversity estimation is equal to:

$$\Theta(k) = \frac{\prod_{i=1}^{k}(NP - i)}{dir \cdot n_1! \cdot n_2!}. \tag{4.4}$$

It is obvious that a high diversity slows down the convergence rate, whereas a low one results either in stagnation or premature convergence. Thus some balance is needed. By controlling the number of randomly extracted individuals (or more precisely a strategy type) we can easily provide the required diversity of population (see Section 3.4).

Practical remark: As you already know, the diversity function depends on the number of individuals used in a strategy and the size of a population. If the number of individuals in a strategy surpasses 7 then the diversity of a strategy becomes enormous and consequently the exploration would be excessive. So, in practice, it is reasonable to use not more than 3, 4, or 5 individuals to give sufficient exploration. This is a practical compromise between computing time and quality of the search space exploration.

4.4 Control Parameters

The goal of control parameters is to keep up the optimal exploration/exploitation balance so that the algorithm will be able to find the global optimum in the minimal time. Practical tests show that within the search process the diversity of a population (its exploration capabilities) usually goes down more rapidly than we would like. Thus, one of the ways to provide good control is to retain the desired diversity level.

4.4.1 Diversity Estimation

During the last years different manners of the diversity estimation were proposed in the literature. I shall represent some of them here.

Expected Population Variance

One of the exploration power measures is the population variance [BD99]

$$\mathrm{Var}(\mathbb{P}) = \overline{X^2} - \overline{X}^2\,, \qquad (4.5)$$

where $\overline{X}$ is a population mean and $\overline{X^2}$ is a quadratic population mean. So, if $\mathrm{Var}(\mathbb{P}^0)$ is an initial population variance, then after several generations the expected population variance can be estimated as a function of the control parameters $\Omega = \Omega(F, Cr, NP, k)$ [Zah01].

$$E(\mathrm{Var}(\mathbb{P}^g)) = \Omega^g \cdot \mathrm{Var}(\mathbb{P}^0)\,. \qquad (4.6)$$

The comparisons of the real and theoretical results confirm the likelihood of such an estimation. To retain the given diversity it is necessary for the transfer function Ω to be a little more than or at least equal to 1: $\Omega \geq 1$.

Average Population Diversity

In the work [Š02] a direct measure of average population diversity was introduced.

$$div(g) = \frac{\sum_{i=1}^{NP} \sum_{j=i+1}^{NP} \frac{|X_i(g) - X_j(g)|}{H - L}}{2 \cdot D \cdot (NP - 1) \cdot NP}\,. \qquad (4.7)$$

It represents for each generation an average normalized distance between the individuals of the population.

Mean Square Diversity

Another direct way to estimate diversity is to use the mean square root evaluation for the population as for its objective function [LL02a].

$$P_{div}^g = \frac{1}{k_p} \sqrt{\frac{1}{NP} \sum_{i=1}^{NP} \sum_{j=1}^{D} (x_{i,j}^g - x_{i,j}^{g-1})^2}$$

$$F_{div}^g = \frac{1}{k_f} \sqrt{\frac{1}{NP} \sum_{i=1}^{NP} (f_i^g - f_i^{g-1})^2}\,, \qquad (4.8)$$

where k_p, k_f compress P_{div}^g, F_{div}^g into the interval $[0, 1]$. This method requires an additional memory source both for the population and the vector of objective functions of the previous generation.

P-Measure

There is a simpler and, perhaps, more practical way to estimate population diversity (see Chapter 5). $P(population)$-measure is a radius of population, that is, an Euclidean distance between the center of population O_p and the farthest individual from it.

$$P_m = \max \|X_i - O_p\|_E, \qquad i = 1, \ldots, NP. \tag{4.9}$$

4.4.2 Influence of Control Parameters

Constant of Differentiation

The constant of differentiation F is a scaling factor of the difference vector δ. F has considerable influence on exploration: small values of F lead to premature convergence, and high values slow down the search. I have been enlarging the range of F to the new limits $F \in (-1, 0) \cup (0, 1+]$ (see Subsection 3.2.5). Usually, F is fixed during the search process. However, there are some attempts to relax this parameter. Relaxation significantly raises the covering of the search space and also partially delivers us from the exact choice of F. Among the relaxations we can outline $F = N(0, F)$, $N(F, \sigma)|_{\sigma \ll F}$, and $N(F, F)$ with a normally distributed step length and the same variants with uniform distribution.

Constant of Crossover

The constant of crossover reflects the probability with which the trial individual inherits the actual individual's genes. Although using Crossover makes the algorithm rotationally dependent (Appendix D and [Sal96, Pri99]), crossover becomes desired when we know the properties of an objective function. For example, for symmetric and separable functions $Cr \approx 1 - 1/D$ is the best choice; for the unimodal (or quasi-convex) functions a good choice is crossover with the best individual. Moreover, small values of Cr increase the diversity of population. To put it differently, the number of potential solutions will be multiplied by the number of vertices of a D-dimensional hypercube built on the trial and target individuals.

Size of Population

The size of population NP is a very important factor. It should not be too small in order to avoid stagnation and to provide sufficient exploration. The increase of NP induces the increase of a number of function evaluations; that is, it retards convergence. Furthermore, the correlation between NP and F may be observed. It is intuitively clear that a large NP requires a small F; that is, the larger the size of a population is, the more densely the individuals fill the search space, so less amplitude of their movements is needed.

Type of Strategy

The strategy can be characterized by the number of randomly extracted individuals k and the dynamic effect resulting from the manner of their use. k controls the diversity, whereas the way to calculate the trial individual directly reflects the dynamics of exploration. A small k makes the strategy a slack one. A big k slows down the convergence rate because of both the excessive diversity and towering complexity of differentiation.

4.4.3 Tuning of Control Parameters

The effective use of an algorithm requires the tuning of control parameters. And this is a time-consuming task. However, the parameter tuning may be replaced by the parameter control [EHM99].

Three types of parameter control are distinguished.

1. *Deterministic control:* parameters are followed by a predefined deterministic law; there is no feedback information from the search process.
2. *Adaptive control:* parameters depend on feedback information.
3. *Self-adaptive control:* parameters depend on the algorithm itself; they are encoded into it.

Deterministic Control

For the first time the deterministic control of the population size has been introduced using the energetic selection principle (Chapter 8). The population is initialized by a huge number of individuals; then an energetic barrier (deterministic function that depends on the generation number) is applied to reduce the population to a normal size. This method leads to global convergence and increases its rate.

Next, the law switching from one type of strategy to another can be implemented. In such a way both the number and type of used individuals are controlled. Switching results from a predefined switch-criterion that depends, for instance, on the relative difference $(f_{max} - f_{min})$.

Adaptive Control

Two methods of adaptive control are distinguished.

1. Refresh of population
2. Parameter adaptation

The refresh of population [Š02] is realized either by replacement of "bad" individuals or by injecting individuals into the population. Both methods increase the diversity. The refresh can be aimed at exploration of new regions of the search space as well as for convergence rate improvement. It repeats periodically or each time when the population diversity reaches a critical level.

The parameter adaptation entirely obeys the state of population. The feedback information calculated on the basis of this state modifies the control parameter according to a control law.

By now, two variants of adaptation have been proposed.

- The first one is a fuzzy control that adjusts the constant of differentiation F. The input signal for the fuzzy system is computed from (4.8). Membership functions and fuzzy rules are established based on expert knowledge and previous tests. Notice that diversity evaluation, fuzzification/defuzzification, and execution of fuzzy rules are time-consuming operations and their complexity might be comparable with one DE generation. Thus, it is always necessary to estimate the relative efficiency of this method.
- The second one is an adaptation based on the theoretical formula of the expected population variance (4.6) [Zah02]. Also, the parameter F is adjusted. But this parameter becomes unique for each gene of the individual; that is, each gene has its own parameter value. The adaptation happens each generation. It is less complex than the previous one ($O(NP \cdot D)$) and does not modify the complexity order of one generation. Such an adaptation prevents premature convergence, but does not ensure the best convergence rate. Moreover, it does not depend on the objective function, so the introduction of supplementary information would be desirable.

Self-Adaptive Control

The work [Abb02] first proposed self-adaptive crossover and differential mutation. Also, separate parameters were proposed for each individual as well as for differential mutation as for crossover. The self-adaptation scheme repeats the principle of differential mutation: $r = r_{\xi_3} + N(0,1) \cdot (r_{\xi_2} - r_{\xi_1})$, $r \in \{F, Cr\}$. The given adaptation was used for multiobjective Pareto optimization and the obtained results outperform a range of state-of-the-art approaches.

4.5 On Convergence Increasing

Actually, we emphasize three trends of convergence improvement.

1. Localization of global optimum
2. Use of approximation techniques
3. Hybridization with local methods

All these trends represent a pure exploitation of available information. They elevate the intelligence of the algorithm in order to improve its convergence. The main purpose of the prescribed improvement is an ability to solve large-scale nonlinear optimization problems.

Localization

The energetic selection principle (Chapter 8) is a particular case that illustrates a fast localization of the global optimum. An initialization by a large population helps to reveal promising zones; then a progressive reduction of the population locates the global optimum. At the end, local techniques can be used.

Approximation

The deterministic replacement of "bad" individuals by "good" ones is one of the ideas to ameliorate the convergence. Let the good individuals be created by approximation methods. For example, we construct a convex function regression on the basis of the best individuals of the population. Then, the optimum of this regression will replace the worst individual. Here, there are lots of regression techniques that could be applied. For instance, a more recent and promising one is support vector machines (Chapter 9). The main emphasis is made on choosing an appropriate kernel function, which considerably influences the quality of approximation.

Local Methods

The most traditional idea is to use the population-based heuristics as "multi-starts" for deterministic optimizers. The positive results were demonstrated by hybridizing DE with the L-BFGS method [AT02]. This hybridization proves to be more efficacious for large-scale problems than for small ones.

Problems

4.1. What do we mean by exploitation and exploration when speaking about evolutionary algorithms?

4.2. What advantages and disadvantages has an excessive exploration? And an excessive exploitation?

4.3. It is well known that the role of genetic operators is to control the balance between exploration and exploitation. If you had a choice between mutation and crossover, what would you prefer? And why?

4.4. Choose from Chapter 3 four different strategies (one strategy from each group) and explain how the strategies realize the functions of exploitation and/or exploration.

4.5. What do we mean by the diversity of population? Analyse how the diversity of population changes as exploitation increases?

4.6. Analyse the efficiency of exploration when the constant of differentiation F increases.

4.7. What is the general operator? Could you consider the operation of differentiation as the general operator? Explain your point of view.

4.8. Explain the disruption effect. Does the differentiation operator possess this effect?

4.9. Suppose n individuals in n-dimensional space E^n are linearly dependent vectors. Then, among them, there exists r linearly independent vectors forming the basis in the subspace $E^r \subset E^n$. Let the optimum $Opt \in E^n$ be outside of subspace E^r, that is, there are no decompositions on basis vectors. The DE algorithm implements only differentiation and selection (without crossover). Is the found solution X^* the optimum Opt? Write your arguments and give an explaining sketch.

4.10. What properties should crossover have? Enumerate at least three properties and give an example for each of them.

4.11. How does the exploitation property appear in crossover?

4.12. In which cases does crossover become useless and may be even harmful?

4.13. Due to what does the self-adaptation of difference vector δ occur?

4.14. How, in theory, does one estimate the diversity of population from a combinatorial standpoint? Take, from Chapter 3, any three strategies and calculate the diversity according to the formula (4.4) of Chapter 4.

4.15. Test the strategies selected for problem (4.14) using any test function you have. Analyse the influence $\Theta(k)$ on the precision of the found solutions and the computing time spent to obtain them. Make a written deduction.

4.16. Why are the control parameters necessary?

4.17. What empiric methods for diversity estimation do you know? Enumerate at least four methods and implement one of them for choice.

4.18. Plot experimental curves (diversity from generation) for the strategies chosen in problem (4.14) and the function used in problem (4.15). Explain the obtained results.

4.19. What is the relaxation of F? Are there any advantages of relaxation? Give some examples of relaxation. Does relaxation have some drawbacks?

4.20. Given a test function, the so-called Salomon function,

$$f(X) = -\cos(2\pi\|X\|) + 0.1 \cdot \|X\| + 1\,,$$

$$\|X\| = \sqrt{\sum_{i=1}^{D} x_i^2}\,, \quad -100 \le x_i \le 100\,,$$

$$f(X^*) = 0\,, \quad x_i^* = 0\,, \quad VTR = 1.0 \times 10^{-6}\,.$$

Plot this function for $D = 2$. Make two experiments: the first for the fixed F and the second for the relaxed F. Compare the results and show at least one advantage and one disadvantage of the relaxation.

4.21. In which cases is crossover definitely necessary? Give concrete examples.

4.22. For any problem you choose, initialize the population as described in problem (4.9). Find the optimal solution without using the crossover operation. Then, add the crossover operation and watch whether the new-found optimal solution is changed? Make tests with different values of crossover. Which value of crossover is the best for your case?

4.23. What chances do you take when the population size is too small? Demonstrate on an example the stagnation effect of the algorithm. For the demonstrated example, plot a graph of the time (generations), needed to find the optimal solution (VTR) from the size of the population NP.

4.24. Choose arbitrarily one of four groups of strategies. Test several strategies of this group on either your favorite problem or any known test function. Plot curves of convergence for these strategies. Analyse how the convergence of the algorithm depends on the number of randomly extracted individuals needed for one or another strategy? Did you use, for validity of results, one and the same initial population for all your tests?

4.25. Why do we need to adjust the control parameters?

4.26. What is the difference between tuning of control parameters and the parameter control? What three types of parameter control do you know?

4.27. Think out your own method of deterministic control, implement it and test. Estimate its efficiency.

4.28. What kinds of adaptive control do you know?

4.29. Elaborate your own method of adaptive control for one of four parameters (F, Cr, NP, k). Implement it and test. Estimate its efficiency.

4.30. Explain in short what is the self-adaptive control.

4.31. Think out your own version of self-adaptive control. Test it and estimate how efficient your version is.

4.32. For optimization of a test function, the so-called Schwefel's function,

$$f(X) = -\frac{1}{D} \sum_{i=1}^{D} x_i \cdot \sin(\sqrt{|x_i|}) , \qquad -500 \le x_i \le 500 ,$$

$$f(X^*) = -418.983 , \qquad x_i^* = 420.968746 , \qquad VTR = 0.01 ,$$

use the DE algorithm with three control parameters (F, Cr, NP). Your task is to optimize these parameters in order to ameliorate the convergence of the algorithm. For this, use an algorithm of global optimization (either DE or another). Plot the Schwefel's function for $D = 2$. Write the program and find the optimal control parameters.

4.33. What three trends of convergence improvement do you know? To what are these improvements usually attributed? Write a short explanatory paragraph.

5

New Performance Measures

In this chapter three new measures of the algorithm's performance are introduced. These measures provide a more objective vision of the behavior of an algorithm and supplement the ensemble of standard measures for evolutionary algorithms. These are the following measures.

1. *Q-measure:* an integral measure that combines the convergence of an algorithm with its probability to converge.
2. *P-measure:* a measure of the population dynamics. It analyzes the convergence from the point of view of a population.
3. *R-measure:* a measure that permits estimating the robustness of a strategy.

5.1 Quality Measure (*Q-Measure*)

Quality measure or simply *Q-measure* is an empirical measure of the algorithm's convergence. It serves to compare the objective function convergence of different evolutionary methods. In the case under consideration it is used to study the strategy behavior of the DE algorithm.

Firstly, we introduce some necessary notation:

- g_{max} — maximal number of generations
- E_{max} — maximal number of function evaluations: $E_{max} = g_{max} \cdot NP$
- n_t — number of trials, launchings of algorithm[1]
- E_i — number of function evaluations in ith trial, $i = 1, \ldots, n_t$
- ε — precision of the solution

---

[1] The core of the algorithm is a random process; thus to obtain an average statistics it is run several times.

Convergence Measure

Suppose, if in the ith trial the given precision ε is not achieved after $E_{\max}$ function evaluations; it is then considered that this trial is not successful. Let there be n_c successful of n_t total trials, where the strategy arrives at the optimal solution with a sufficient precision $\| x - x^* \| \leq \varepsilon$, so the convergence measure is calculated as

$$C = \frac{\sum_{j=1}^{n_c} E_j}{n_c}, \tag{5.1}$$

where $j = 1, \ldots, n_c$ are successful trials. In other words, the convergence measure is an average number of function evaluations for successful trials.

Probability of Convergence

It measures the probability of convergence and can be calculated as the percentage of successful to total trials.

$$P_C = \frac{n_c}{n_t} \% . \tag{5.2}$$

Q-measure as an Integrating Criterion

Q-measure[2] incorporates the two criteria mentioned above. By combining the convergence rate and its probability in one, the Q-measure undoubtedly destroys some data about algorithm performance. Nevertheless we can still refer to C and P_C if we need them. Now Q-measure is a single criterion to be minimized. Without Q-measure the algorithm performance would be ambiguous (as in multiobjective optimization). The formula of Q-measure is

$$Q_m = \frac{C}{P_C} . \tag{5.3}$$

Enes and Q-Measure (Advantage of Q-Measure)

The average number of function evaluations' overall trials can also be calculated.

$$Mean = \frac{\sum_{i=1}^{n_t} E_i}{n_t} \tag{5.4}$$

and *Enes*-measure, introduced by Price in [Pri97]

$$Enes = \frac{Mean}{P_C} = \frac{\sum_{i=1}^{n_t} E_i}{n_c} . \tag{5.5}$$

When all the trials are successful $Q_m = C = Mean = Enes$, but as P_C decreases, Q_m increases. This way, a lower convergence probability and a higher number of function evaluations both increase Q_m. *Enes* gives very similar results when $E_{\max}$ is the same order of magnitude as C, but the main advantage of Q_m is that it does not depend on an arbitrary $E_{\max}$. The Q-measure function is shown in Fig. 5.1.

---

[2] Q-measure was introduced during personal communications with Kenneth Price.

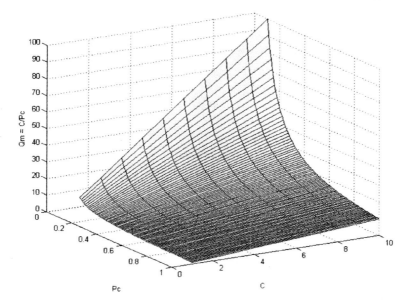

Fig. 5.1. Q-measure criterion function.

5.2 Entropy

Based on the concept of entropy Price introduced a new measure to estimate the complexity of an algorithm tuning. This is *crafting effort* [Pri97], that is, the effort needed to find a good set of control parameters. We use this criterion to compare an effort of a strategy tested on benchmarks. So, it is possible to tell if that strategy is easier to tune in comparison with this one.

Let there be n_f test functions and three control parameters $cp_i : \forall i \in \{1, 2, 3\}$, where the size of the population is $NP = cp_1$, the constant of differentiation is $F = cp_2$, and the constant of crossover is $Cr = cp_3$. For each test function the optimal control parameters' values are found from a fixed value grid $\{v_{cp_i}^k\}_{k=1}^{n_{cp_i}}$. So, we have three vectors of n_f optimal values for each control parameter separately, $V_{cp_i}^*$. Each value in these vectors takes place $n_j : 0 \le n_j \le n_f$ times, so that $\sum_j n_j = n_f$. Thus, crafting effort of the control parameter is

$$\vartheta_{cp_i} = -\sum_j \frac{n_j}{n_f} \cdot \log \frac{n_j}{n_f} \,, \tag{5.6}$$

and total crafting effort of the strategy is

$$\Theta = \sum_{i=1}^{3} \vartheta_{cp_i} \,. \tag{5.7}$$

On the other hand, the total crafting effort calculated for a strategy is inversely proportional to its robustness: the less crafting effort, the more robust the strategy is; that is, the same set of control parameters gives a good performance for most objective functions,

$$robustness \sim \frac{1}{\Theta} .$$

(5.8)

5.3 Robustness Measure (R-$Measure$)

I propose another approach to calculate the robustness. This approach is based on statistics.

Suppose that for each strategy and for each n_f test function the best combinations of the control parameters $\{F^*, Cr^*, \eta^*\}$ are chosen from a given grid of value. $\eta^* = NP^*/D$, where D is a dimension of the variables' space. Then, the standard deviation of these parameters is calculated separately:

$$\sigma_X = std\{X_j^*\}_{j=1}^{n_f}, \qquad X \in \{F, Cr, \eta\} .$$

(5.9)

It is assumed that the inverse of dispersion of these deviations represents the robustness of a strategy with respect to each of the control parameters.

In order to have a single criterion we normalize the deviations as $\sigma_X/\triangle_X$, where $\triangle_X$ is a range of parameter changing. So, these three values can be drawn as edges of the orthogonal parallelepiped in 3-D space. The volume of such a parallelepiped integrates these three deviations in one. The inverse of this volume characterizes the robustness of the strategy. I call it the R-$measure$. Thus, the formula of the R-measure is:

$$R_m = \frac{\triangle_F}{\sigma_F} \cdot \frac{\triangle_{Cr}}{\sigma_{Cr}} \cdot \frac{\triangle_\eta}{\sigma_\eta} .$$

(5.10)

5.4 Population Convergence (P-$Measure$)

Let us imagine a is a known optimum that minimizes an objective function $\min f(x) = f(a) = L$. So, from an optimization point of view it is necessary to find a solution x^o that minimizes $f(x)$ with a given precision $\varepsilon : |f(x^o) - L| < \varepsilon$. Consequently, there exists some $\delta : 0 < \|x^o - a\| < \delta$ that represents the distance between the optimum a and the found solution x^o.

Convergence of an Objective Function

When we speak about algorithm convergence, usually we mean the convergence of the objective function that we minimize. The rate of convergence is generally described by a functional dependence, $f(gen)$, where gen is the current generation. It looks like Fig. 5.2, for example.

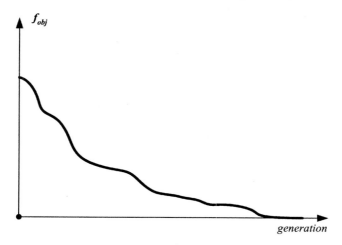

Fig. 5.2. Convergence of an objective function.

Convergence of a Population

By analogy with limits, we introduce a measure of the convergence of a population. The convergence of a population is a measure that allows us to observe the convergence rate from the point of view of variables. It represents the dynamics of grouping individuals around the optimum. In general, it is supposed that the more densely individuals are located, the better algorithm convergence is intended. To measure this characteristic I have introduced the radius of population gauging. I called it population measure or simply *P-measure*. P-measure is a radius of a population, that is, the Euclidean distance between the center of a population and the individual farthest from it. The center of a population is calculated as the average vector of all individuals (barycenter) $O_p = \sum_{i=1}^{NP} ind_i / NP$. Thus, the P-measure is defined as

$$P_m = \max \| ind_i - O_p \|_E, \qquad i = 1, \ldots, NP. \tag{5.11}$$

The population radius measure (P-measure) expands the convergence measure of an objective function. And so, in the aggregate, both measures give a better convergence image. I shall show them in the same figure, Fig. 5.3.

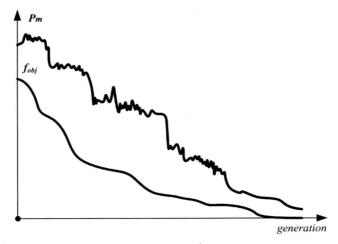

Fig. 5.3. Convergence of a population.

Problems

5.1. What standard performance measures do you know? Which of them have you already used to estimate the efficiency of an algorithm? How objective an estimation did they provide?

5.2. Why do we compare algorithms using average statistical results, starting each algorithm being compared several times?

5.3. How does one estimate the convergence of an algorithm? What is the probability of convergence?

5.4. What is the Q-measure? What are its advantages and disadvantages in comparison with the measures cited in problem (5.3)? Explain how Q-measures differ from Enes-measures?

5.5. Given a test function, the so-called Griewank's function,

$$f(X) = \frac{1}{4000} \cdot \sum_{i=1}^{D} x_i^2 - \prod_{i=1}^{D} \cos\left(\frac{x_i}{\sqrt{i}}\right) + 1$$

$$-600 \leq x_i \leq 600$$

$$f(X^*) = 0, \quad x_i^* = 0, \quad VTR = 1.0 \times 10^{-6}.$$

Plot this function and its level lines for $D = 2$. Write the code to calculate the Q-measure. Using two different sets of control parameters, run the DE algorithm 50 times for each of the cases. Calculate the Q-measure for each set of parameters. Which of the two control parameters sets is more appropriate for solving this function?

5.6. What does "crafting effort" mean? Explain the formula of crafting effort for one control parameter, F or Cr for choice.

5.7. Give a definition of "robustness". How, in a qualitative sense, is crafting effort related to the robustness of a strategy? Write a short explanation with an example.

5.8. Calculate R-measure for two different strategies taken from Chapter 3. For this, use any five test functions or optimization problems. Define which of two strategies is more robust.

5.9. Why do we need to trace the convergence of population? Give an example when it is definitely necessary.

5.10. Given a test function, the so-called Bohachevsky's function

$$f(X) = x_1^2 + 2x_2^2 - 0.3\cos(3\pi x_1) - 0.4\cos(4\pi x_2) + 0.7$$

$$-50 \leq x_1, x_2 \leq 50$$

$$f(X^*) = 0, \quad X^* = 0, \quad VTR = 1.0 \times 10^{-6}.$$

How does one estimate empirically the convergence of population? Propose your own method. Implement both methods and plot graphs both of the objective function's convergence and the convergence of population for each method. What kind of information can you extract from these curves?

5.11. Think out your measures to estimate the algorithm's performance.

6

Transversal Differential Evolution

In this chapter I introduce the new concept of evolution: each individual, before returning into the population, makes several steps (a trajectory) in a space, and only then it comes back to the population. The research in this direction allowed me to allot three evolutionary species of an algorithm. These are: two-array species, sequential species, and, the new (proposed by me) transversal one. The two-array species is the oldest species adopted by the evolutionary computation community. It is designed mainly for synchronous parallel implementation. Next, sequential species is intuitively used by many people and has the following advantages: reasonable memory requirements, instantaneous renovation of a population, and better convergence. The transversal species constitutes the generalization of the sequential one. This technique permits the interpretation of some other recent metaheuristics, for example, free search. On the other hand, the transversal species is very well adapted to an asynchronous parallel implementation of the algorithm. And moreover, the number of transversal steps permits controlling the diversity of a population in a delicate manner.

6.1 Species of Differential Evolution

In my previous works I usually used the algorithm of differential evolution stated in Chapter 2 (Section 2.3). However, the first species of DE proposed by R. Storn and K. Price [SP95] was realized, with the parallelization purposes, as two population arrays. Now, I shall introduce a new kind of DE, which I called *transversal* differential evolution. The next facts served me as the main motivation for the introduction of such an algorithm:

1. Many papers devoted to studying the influence of population diversity and of the exploitation/exploration trade-off on the convergence rate (see Chapter 4).

2. Aspirations to continue the universalization of DE; that is, indirectly speaking, I wished that DE would completely generalize and involve a recently proposed Free Search method [Pen04].
3. Also new parallelization technologies for heterogeneous networks (*mpC*) brought me to a transversal architecture of the algorithm [Las03].

Let me briefly present the existing, two-array and sequential, DE species and then we shall examine together the new transversal one.

6.2 Two-Array Differential Evolution

This species of DE uses two populations: the old and the new one. During each iteration cycle (evolutionary step) individuals of the old population are transferred into the new population by reproduction and selection operations. In order to generate trial individuals the old population is always used, that is, all individuals necessary for differentiation are randomly extracted from the old population. After crossover and selection the obtained individual passes into the new population. At the end of the iteration the new population becomes the old one and then the next evolutionary step begins (see Fig. 6.1).

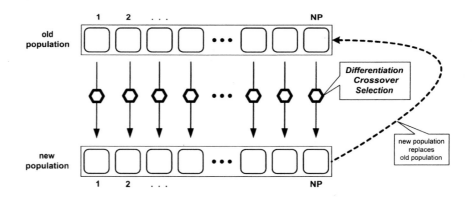

Fig. 6.1. Two-array differential evolution.

Such an organization allows us to treat each individual independently and leads to synchronous parallelization. We shall call this algorithm *two-array* DE, because two arrays are used to operate with a population. We show the changes in the classical algorithm (see Alg. 4).

Algorithm 4 *Two-array* Differential Evolution

Require: $\cdots$, $\mathbb{P}^0_{old}$, $\mathbb{P}^0_{new}$ – 2 arrays
   initialize $\mathbb{P}^0_{old} \leftarrow \{ind_1, \dots, ind_{NP}\}$
   evaluate $f(\mathbb{P}^0_{old})$
   while (not terminal condition) **do**
      for all $ind \in \mathbb{P}^g_{old}$ **do**
         $\mathbb{P}^g_{old} \rightarrow \pi = \{\xi_1, \xi_2, \dots, \xi_n\}$
         $\cdots\cdots\cdots\cdots$
         $ind \rightarrow \mathbb{P}^g_{new}$
      end for
      $\mathbb{P}^{g+1}_{old} \leftarrow \mathbb{P}^g_{new}$
      $\cdots\cdots$
   end while

Remark

Using two populations is a memory-consuming approach, especially for large-scale problems. The fact that during the whole iteration we operate only with the old population brings the diversity to a constant level within this iteration.

6.3 Sequential Differential Evolution

The algorithm of sequential DE is described in Chapter 2 (see Alg. 3). In this algorithm one population array is implemented. Individuals evolve one after another (from 1 to NP) and immediately return in the population. The order of evolving is not fixed, but each individual evolves once per iteration. Figure 6.2 shows this principle.

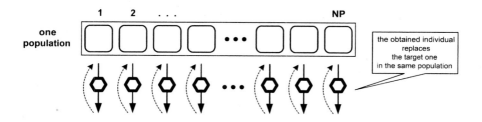

Fig. 6.2. Sequential differential evolution.

Remark

By using one array to keep up the population we considerably reduce memory needs. Also, previously improved individuals at this evolutionary step immediately participate in creation of trials (differentiation). In such a way each successive individual will be created on the basis of better individuals. Thus, in this case, the diversity of a population decreases proportionally to approaching the optimum.

6.4 Transversal Differential Evolution

Now we allow an individual to evolve several (say n) times before it replaces the target one in the population. Graphically, such an individual passes a nonlinear trajectory framed by reproduction operators (n jumps on the search space) and it chooses the best location. Then, we proceed with the next individual.

This behavior describes an intermediate stage between two previous DE species. The population keeps up the constant diversity level during these n steps. The trajectory represents a new subpopulation constructed on the basis of individuals of the population. On the other hand, by selection, the best individual of this new subpopulation immediately replaces the target one in the initial population. And furthermore, the obtained individual is used to form posterior individuals of the population.

In order to draw an analogy we could imagine two-array DE where the "new" population consists of n individuals and is rotated transversely to the "old" one. This "new" population is sequentially created for each individual of the "old" population. The individuals evolve into the "new" population without the selection operation. And at the end, the best individual of the "new" population replaces the target individual in the "old" population. If $n = 1$ (one step) then transversal DE turns into a sequential one. But there is no such transformation into two-array DE.

We shall baptize the presented algorithm *transversal differential evolution*. This term explains the algorithmic architecture well: in addition to the consecutive ("longitudinal") transformation of a population (sequential DE, see Fig. 6.2), at first, n "transversal" steps are applied to each individual (see Fig. 6.3). The algorithm modification is the following (see Alg. 5).

Algorithm 5 *Transversal* Differential Evolution

$\cdots\cdots$
    while (not terminal condition) **do**
      for all $ind \in \mathbb{P}^g$ **do**
        for transversal step $= 1$ to n **do**
          $\cdots$ $ind \leftarrow \{\text{Diff, Cross, Select}\}$
        end for
      end for
      $\cdots$
    end while

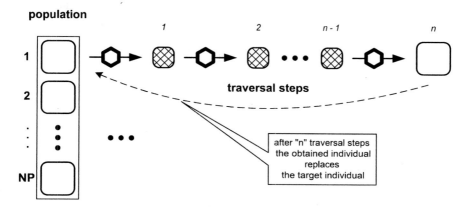

Fig. 6.3. Transversal differential evolution.

Transversal DE may be associated with a free search algorithm [Pen04]. Free search (FS) has been recently created on ideas inspired by animal behavior as well as particle swarm optimization and differential evolution. Its main principle is to give excessive freedom for the search space exploration.

A new FS individual is formed by the relaxed strategy $\omega = \xi + F_{rel} \cdot (H - L)$, $F_{rel} = F \cdot rand[0, 1]$. However, as shown for DE [RL03], this F relaxation does not do the essential convergence improvement. Moreover, the relaxation within boundary constraints does not permit us to control a needed level of the population diversity. So, a wide range of the strategies proposed for DE (2.8) will provide undoubtedly more interesting results, Although there were no comparative results until now. Nevertheless, we can clearly observe that transversal DE is ideally close to FS and can involve FS as one of its strategies.

6.5 Experimental Investigations

In this section we shall compare the efficiency of three outlined DE species. In order to do this we chose some well-known benchmarks (see Table 6.1 and Appendix C). In this table VTR (value to reach) defines the desired proximity to the optimum and serves as a terminal condition. IPR sets the boundary constraints: $X \in [-IPR, +IPR]$. D and NP are the dimension and population size of a tested problem, accordingly. The constants of differentiation and of crossover are fixed as well as the strategy type ($F = 0.5$, $Cr = 0$, $\omega = \xi_3 + F \cdot (\xi_2 - \xi_1)$). For statistical purposes I ran each test 20 times and then found the average values. Moreover, the same initial populations are used to test different DE species.

First, we shall study the behavior of transversal DE relative to the sequential one. We begin to increase the number of transversal steps from 1 to

Table 6.1. Benchmarks for Evolutionary Algorithms

Function Name	VTR	IPR	D	NP
f_1 – Ackley's function	1e–3	32.768	30	300
f_2 – Rotated ellipsoid	1e–6	65.536	20	200
f_3 – Rosenbrock's function (Banana)	1e–6	2.048	10	200
f_4 – Step function	1e–6	5.12	5	100
f_5 – Shekel's function (Foxholes)	0.998005	65.536	2	40

30 and observe the changes in convergence rate (number of function evaluations). Note that transversal DE with one transversal step will be equivalent to the sequential one. The obtained results are summarized in Table 6.2. It is obvious that sequential DE and/or a very few number of transversal steps give, in most cases, good results. It can be explained by optimal population diversity decreasing. In the cases with specific functions, where extra diversity is needed, slight increasing of the number of transversal steps (up to 5–10% of NP) leads to better results. Thus, the transversality of DE may be presented also as one of the ways of controlling diversity.

Table 6.2. Transversal Differential Evolution

Number of	Number of Function Evaluations				
Transversal	f_1	f_2	f_3	f_4	f_5
Steps	Ackley	Ellipsoid	Banana	Step	Foxholes
1	**266362,60**	**105414,60**	66695,60	15714,30	1936,80
2	280021,40	107914,10	67669,00	15921,70	**1936,30**
3	273918,10	105874,00	**65945,10**	16368,80	1970,30
4	277127,80	109447,70	69489,20	17523,60	1956,00
5	277109,90	110563,90	71597,50	16925,30	2059,30
10	305529,30	119798,20	81687,30	**14727,90**	2133,60
20	342003,30	145450,00	98967,80	15916,60	2426,20
30	381638,00	172979,10	116217,90	20228,80	—

Second, we shall compare two-array DE implementation with sequential DE and 10-transversal DE (10 transversal steps). The results are shown in Table 6.3. This comparison illustrates the relative positions of each species. In general, we can see that the most efficient is sequential DE; then, in several cases, transversal DE is comparable with the sequential one, and at the end there is two-array DE.

The experimental study confirms the theoretical premises stated in Section 6.3.

> **Diversity of a population should decrease proportionally to approaching the optimum.**

Table 6.3. Comparison of Differential Evolution Species

DE Species	Number of Function Evaluations				
	f_1 Ackley	f_2 Ellipsoid	f_3 Banana	f_4 Step	f_5 Foxholes
Sequential DE	266362,60	105414,60	66695,60	15714,30	1936,80
10-transv. DE	305529,30	119798,20	81687,30	14727,90	2133,60
Two-array DE	301104,25	123007,05	88095,05	17120,80	2365,93

Transversal DE plays an intermediate role between sequential and two-array ones.

After this study the positive effect of precalculated differentials in topographical DE [AT02] can be well explained by the needs of population diversity keeping, whereas the gradient methods reduce it considerably. In other cases, an excessive diversity or diversity keeping could not be recommended.

6.6 Heterogeneous Networks of Computers

Recently heterogeneous networks of computers become the most used for parallel computation. It is explained by their high productivity at very low cost. The main principle is to use many personal computers with different capacities integrated in a common network to solve a complex task. The portions of the task are divided among computers proportionally to their capacities (performances of processors and links). There are many parallel languages for programming on heterogeneous networks. One of them, the most recent and advanced, is mpC (multiparallel C). It is an extension of the ANSI C language, designed specially to develop portable adaptable applications, possessing intelligent parallel-debugging tools (see [Las03] and http://www.ispras.ru/~mpc).

mpC[1] allows programmers to implement their heterogeneous parallel algorithms by using high-level language abstractions rather than going into details of the message-passing programming model of the MPI level. Moreover, it handles the optimal mapping of the algorithm to the computers of the executing heterogeneous network. This mapping is performed at runtime by the mpC programming system and is based on two performance models:

1. Performance model of the executing heterogeneous network
2. Performance model of the implemented algorithm

The performance model of the heterogeneous network of computers is summarized as follows.

---

[1] Specifications of mpC were generously given by A. Lastovetsky, the author of the programming language.

- The performance of each processor is characterized by the execution time of the same serial code.
- The communication model is seen as hierarchy of homogeneous communication layers. Each is characterized by the latency, overhead, and bandwidth. Unlike the performance model of processors, the communication model is static: its parameters are obtained once at the initialization of the environment.

The performance model of the implemented algorithm is provided by the application programmer and is a part of the mpC application. The model is specified in a generic form and includes:

- The number of processes executing the algorithm
- The total volume of computation to be performed by each process during the execution of the algorithm
- The total volume of data transferred between each pair of the processes during the execution of the algorithm
- How exactly the processes interact during the execution of the algorithm; that is, how the processes perform the computations and communications (which computations are performed in parallel, which are serialized, which computations and communication overlap, etc.)

The mpC compiler will translate this model specification into the code calculating the total execution time of the algorithm for every mapping of the processes of the application to the computers of the heterogeneous network. In the mpC program, the programmer can specify all parameters of the algorithm. In this case, the mpC programming system will try to find the mapping of the fully specified algorithm that minimizes its estimated execution. At the same time, the programmer can leave some parameters of the algorithm unspecified (e.g., the total number of processes executing the algorithm can be unspecified). In that case, the mpC programming system tries to find both the optimal value of unspecified parameters and the optimal mapping of the fully specified algorithm.

Suppose that we want to parallelize two-array DE. So at each iteration we are obliged to synchronize the processes to refresh the old population by the new one. Besides, executing one individual at the processor would be reasonable only with a very complex (time-consuming) objective function. Thus it leads to loss of time and applying heterogeneous networks would be irrelevant.

Let us look at transversal DE. This algorithm permits increasing the amount of computations at the processor. Now a processor deals with entire transversal computations. So, all individuals are treated in parallel. And as the transversal line is finished, the new (best) individual returns to a population. In this case, we don't worry that new individuals would refresh the population simultaneously. Once the target individual is replaced by a better one, it immediately participates in the creation of new trials.

Such an organization leads to asynchronous realization and saves computing time. Furthermore, as theoretical inferences and experimental calculations show, the given species increase the convergence rate. So, transversal DE may be highly recommended for execution on heterogeneous networks of computers instead of the usually used two-array DE because it provides both the flexibility and quality of parallelization.

Problems

6.1. What three species of differential evolution do you know?

6.2. Explain the architecture of two-array differential evolution. What is it used for? Do you see any drawbacks of two-array species?

6.3. Program the two-array species modifying the algorithm described in Chapter 1. Compare the results obtained before and after the modification using, for example, the following test function, the so-called Aluffi-Pentini's problem,

$$f(X) = 0.25x_1^4 - 0.5x_1^2 + 0.1x_1 + 0.5x_2^2$$
$$-10 \le x_1, x_2 \le 10$$
$$f(X^*) \approx -0.3523 , \quad X^* = (-1.0465, 0) .$$

6.4. In which cases should you give preference to a synchronous parallelization? Give a real example.

6.5. What is the difference between sequential and two-array differential evolution? List the advantages of sequential species.

6.6. When does one shows preference to an asynchronous parallelization? Give a real example.

6.7. Describe the framework of transversal differential evolution. What are transversal steps?

6.8. Show that the sequential DE is a special case of the transversal one.

6.9. Write the algorithm for transversal differential evolution and then test all three species (two-array, sequential and transversal) of DE on problems (2.24) and (2.26) of Chapter 2. Justify the obtained results.

6.10. Experiment with the choice of the number of transversal steps. How does it influence the convergence of the algorithm.

6.11. How does the number of transversal steps influence the population diversity?

6.12. What is the benefit of heterogeneous networks of computers?

6.13. What languages for programming on heterogeneous networks of computers do you know?

6.14. Propose your own way of parallelization of transversal differential evolution on heterogeneous networks of computers. Build the action diagram of your method.

7

On Analogy with Some Other Algorithms

Nothing tempts a person as much as a cure for all problems. Nothing delights the scientist as much as a universal algorithm capable of efficiently solving any problem. We all well know that there is no such "remedy", but we are in continuous expectation of its appearance. Differential evolution seems to be a gleam of hope. In this chapter we shall establish an analogy between differential evolution and some other popular algorithms. It is obvious that DE can be easily compared with genetic algorithms and evolution strategies; I leave this task to the mercy of the reader. But here, we shall compare DE with nonlinear simplex, a very old and famous algorithm, and with two recent and efficient metaheuristics, particle swarm optimization and free search. From the outside, drawing an analogy will help us to disclose advantages and disadvantages of differential evolution. *Ex altera parte*, on the inside, I shall make an attempt to interpret the other algorithms through Differential Evolution.

Direct search methods, the methods we are speaking about in this book, firstly were proposed in the 1950s. The state of the art at that time was presented by Swann in 1972 [Swa72]. These methods, as you know, are used in one or more of the following cases.

1. Calculation of the objective function is time-consuming.
2. The gradient of the objective function does not exist or it cannot be calculated exactly.
3. Numerical approximation of the gradient is slow.
4. Values of the objective function are "noisy".

In order to disengage oneself from constraint-handling techniques and to devote one's attention to the algorithms themselves we shall consider only boundary constraints (2.15).

7.1 Nonlinear Simplex

The class of simplex direct search methods was introduced in 1962 [SHH62]. The most famous direct search method was suggested by Nelder and Mead in 1965 [NM65]. I shall briefly describe this method [Wri95].

Four operations, characterized by scalar parameters, are defined: *reflection* (ρ), *expansion* (χ), *contraction* (γ), and *shrinkage* (σ). In the original version of the algorithm these parameters should satisfy

$$\rho > 0, \quad \chi > 1, \quad \chi > \rho, \quad 0 < \gamma < 1 \quad \text{and} \quad 0 < \sigma < 1. \tag{7.1}$$

The standard version of the algorithm assumes that

$$\rho = 1, \quad \chi = 2, \quad \gamma = 1/2 \quad \text{and} \quad \sigma = 1/2. \tag{7.2}$$

The Nelder–Mead algorithm is:

1. **Order.** Order the $D + 1$ vertices in $\mathbb{R}^D$ space to satisfy

$$f(x_1) \leq f(x_2) \leq \cdots \leq f(x_{D+1}). \tag{7.3}$$

2. **Reflect.** Find the *reflection point* x_r from

$$x_r = \bar{x} + \rho \cdot (\bar{x} - x_{D+1}), \tag{7.4}$$

where $\bar{x} = \sum_{i=1}^{D} x_i / D$. Evaluate $f_r = f(x_r)$.
If $f_1 \leq f_r < f_D$ then accept the reflected point x_r and terminate the iteration.

3. **Expand.** If $f_r < f_1$ then calculate the *expansion point* x_e,

$$x_e = \bar{x} + \chi \cdot (x_r - \bar{x}), \tag{7.5}$$

and evaluate $f_e = f(x_e)$. If $f_e < f_r$ then accept x_e and terminate the iteration, otherwise accept x_r and terminate the iteration.

4. **Contract.** If $f_r \geq f_D$ then perform a *contraction* between the better of x_{D+1} and x_r.
   (a) **Outside.** If $f_D \leq f_r < f_{D+1}$ then *outside contraction*

$$x_c = \bar{x} + \gamma \cdot (x_r - \bar{x}), \tag{7.6}$$

and evaluate $f_c = f(x_c)$. If $f_c \leq f_r$ then accept x_c and terminate the iteration, otherwise perform a *shrink*.
   (b) **Inside.** If $f_r \geq f_{D+1}$ then perform an *inside contraction*

$$x_{cc} = \bar{x} - \gamma \cdot (\bar{x} - x_{D+1}), \tag{7.7}$$

and evaluate $f_{cc} = f(x_{cc})$. If $f_{cc} < f_{D+1}$ then accept x_{cc} and terminate the iteration, otherwise perform a *shrink*.

5. **Shrink.** Evaluate f at the D points for the next iteration from

$$v_i = x_1 + \sigma \cdot (x_i - x_1), \qquad i = 2, \ldots, D+1 . \tag{7.8}$$

In spite of world-wide popularity this algorithm suffers from the following drawbacks [Tor89, Wri96].

- It fails when the simplex collapses into a subspace, or becomes extremely elongated and distorted in shape. In most of these cases, the objective function has highly elongated contours and a badly conditioned Hessian.
- It fails when its search direction becomes nearly orthogonal to the gradient.
- It is very sensitive to an increase of the problem dimension.

Now, let us examine a DE strategy from the RAND/DIR group (see Chapter 3). For the dimension D, at each iteration (generation g) we shall randomly extract $D+1$ individuals from the population $\mathbb{P}^g$. The worst individual of this subpopulation (x_{D+1}) belongs to the negative class C_-; the others $(x_i, i = 1, \ldots, D)$ form the positive class C_+. There is no average shift in this strategy. So, the strategy can be rewritten as

$$\omega = \bar{x} + F \cdot (\bar{x} - x_{D+1}) , \tag{7.9}$$

where $\bar{x} = \sum_{i=1}^{D} x_i/D$.

It is obvious that such a strategy is identical to the *reflection* in the Nelder–Mead algorithm (7.4) taking into account that $F \equiv \rho$. Moreover, a wide range of differentiation constant values, $F \in (-1, 2+)$, can be easily interpreted as an *expansion*, $F \geq 2$, and an inside, $F \in (-1, -0)$, or outside, $F \in (+0, 1)$, *contraction*.

Unlike the logic of passing from one step to another in the Nelder–Mead algorithm, differential evolution immediately (or after crossover) selects the best individual among the target and the trial ones. The value of the differentiation constant is either controlled by an adaptation scheme or could be perturbed randomly. Also, the "simplex" is created randomly on the basis of the population for each individual per generation. Furthermore, there is no restriction on the number of used individuals.

I shall emphasize the following advantages/features of the DE approach in comparison with the Nelder–Mead simplex:

- *Search is performed in random subspaces.*
  1. The fact that in strategy (7.9) or, more generally (3.7), any number of individuals (usually $n < D$) can be used illustrates the creation of a simplex in subspaces of the search space.
     This makes the algorithm less sensitive to the problem dimension. On the other hand, such a strategy is more flexible in assigning a descent direction.

2. The subspace, generated from a population, better fits the optimal zones within each new generation.
   Thus, even if an inefficient simplex has appeared at the next step there is a great probability of constructing an efficient one. Often, a "bad" simplex executes the role of an explorer by testing unknown regions.
3. Introduced in (3.7), *average shift* allows us to correct a rough direction. In addition, hybridization with the best individual (RAND/BEST/DIR group) localizes, in some cases, the optimum more briskly.

- *Existence of many simultaneous optimizers.*
   DE could be perceived as a set of autocorrelated simplex optimizers. Such a distributed organization provides more thorough exploration, and improves convergence and precision of solution.

So, as you can see, DE overcomes all the drawbacks stated for the Nelder–Mead simplex. The DE strategy (7.9) of the RAND/DIR group inherits and develops the ideas underlying the simplex algorithm.

7.2 Particle Swarm Optimization

The first particle swarm optimization (PSO) method was proposed by J. Kennedy and R.C. Eberhart in 1995 [KE95]. It progressed simultaneously with DE and, at present, possesses one of the best performances among evolutionary algorithms, or even more widely, among metaheuristics. PSO issued from the metaphor of human sociality. It was born of attempts to simulate human cognition and to apply this model to a real optimization problem.

The idea is to use a set of individuals (a swarm of particles) for search space exploration. A particle represents a vector solution $x_i \in \mathbb{R}^D$, $i = 1, \ldots, NP$ of an optimization task (the same notation as for DE is used here). At each iteration t the particle changes the position influenced by its velocity $v_i(t)$.

$$x_i(t) = x_i(t-1) + v_i(t). \tag{7.10}$$

In order to update the velocity two rules are combined.

1. **Simple Nostalgia**
   In the view of psychology it realizes the tendency of an organism to repeat successful behaviors from the past or, in case of failure, to return to the last success. Let p_i be the best solution attained of the ith particle up to the present iteration, thus the velocity is updated in the following way.

$$v_i(t) = v_i(t-1) + \rho_1 \cdot (p_i - x_i(t-1)). \tag{7.11}$$

2. **Social Influence**
   In spite of an infinite number of possibilities to represent a social behavior two methods were distinguished for defining a neighborhood:

- *gbest* — considers the entire population as a neighborhood;
- *lbest* — defines the subpopulation surrounding the particle.

The *gbest* case is more practical and generally gives better results. Let p_g be the best solution of the population, so the mathematical expression of the social influence is $\rho_2 \cdot (p_g - x_i(t-1))$.

In the view of sociology this term represents the tendency of an organism to emulate the success of others.

To sum up, each of the particles is updated per iteration in the following way.

$$
\begin{aligned}
v_i(t) &= v_i(t-1) + \rho_1 \cdot (p_i - x_i(t-1)) + \rho_2 \cdot (p_g - x_i(t-1)) \\
x_i(t) &= x_i(t-1) + v_i(t)\,.
\end{aligned}
\tag{7.12}
$$

The constants ρ_1 and ρ_2 are control parameters. They define which of two behaviors is dominant.

However, this algorithm in such a form has several drawbacks, leading mainly to premature convergence. In order to improve its performance some modifications were proposed:

- Limitation of $\rho_{1,2}$ up to 2 and its relaxation $\rho_{1,2} \cdot rand(0,1]$
- Limitation of the velocity $v_i \in [-V_{\max}, +V_{\max}]$, $V_{\max} = H - L$
- Introduction of the inertia weight w applied to the previous velocity $v_i(t-1)$, which understates the influence of the preceding behaviors on the current one

All this results in the next update formula:

$$
\begin{aligned}
v_i(t) &= w(v_i(t-1)) + \rho_1 \cdot rand(0,1] \cdot (p_i - x_i(t-1)) \\
&\quad + \rho_2 \cdot rand(0,1] \cdot (p_g - x_i(t-1)) \\
v_i &\in [-V_{\max}, +V_{\max}]\,; \quad \rho_1, \rho_1 \in (0,2]\,.
\end{aligned}
\tag{7.13}
$$

Let us compare PSO with DE now. The PSO strategy (7.13) consists of three components:

1. Damped history of previous velocities $w(v_i(t-1))$
2. Personal behavior $\rho_1 \cdot rand(0,1] \cdot (p_i - x_i(t-1))$
3. Social behavior $\rho_2 \cdot rand(0,1] \cdot (p_g - x_i(t-1))$

The history of previous velocities in terms of psychology characterizes the memory of an organism, and the damping mechanism realizes its property — forgetting. Memory always appears together with personal qualities. If the strategy is built only on personal behavior (second component), the algorithm will not work at all. Thus, memory induces positive effects for an individual movement of the particle, and, on the other hand, it retards its social reaction.

DE does not contain the first two aspects of PSO in pure form. Most likely the transversal technique (see Chapter 6) could illustrate personal behavior.

The individual makes a random walk in the search space and then chooses its optimal (best) position. Here, in DE, the walk is defined by the state of the population, whereas in PSO the next position (of the walk) is defined by the personal optimal position p_i and by the particle's velocity v_i. In PSO terms, DE (differentiation) is more social strategy. Nevertheless, I have made an attempt to introduce into DE the memory aspect in the form of damped preceding velocities (difference vectors). Various damping mechanisms, linear and nonlinear, were tested. As a result, such a modification induced only an inertia of search and showed decrease of convergence.

The third aspect of PSO (social behavior) can be interpreted in DE by the following strategy.

$$\omega = V_b + F^* \cdot (V_b - ind) , \qquad F^* = F \cdot rand(0, 1] . \tag{7.14}$$

This strategy is an example of the RAND/BEST group (see Chapter 3, Equation (3.8)). We easily catch an identity between this strategy and the social aspect of PSO ($\omega = x_i(t)$, $V_b = p_g$, $F = \rho_2 - 1$, $ind = x_i(t-1)$).

Finally, two complementarity features might be observed:

1. DE enlarges the social behavior by its groups of strategies.
2. PSO propagates the ideas of leadership on developing a local neighborhood.

It should be noticed that PSO uses the selection operation in an implicit form, whereas DE regards selection as a separate operation.

The success of DE may be explained by its *collective intelligence* behavior. If PSO exploits only two optimal positions (the particle's optimal position and leader's position), DE involves in evolution the positions and fitness of all the individuals of a population (population state). From a sociopsychological point of view, PSO represents the *conscious*, and DE the *unconscious* way of reasoning. It is well known that a human brain treats about only 10% of information consciously and 90% of information unconsciously (or modified states of consciousness). Perhaps, by imitating the human, the best of universal optimizers would be an alternation of PSO and DE with near to natural proportions.

7.3 Free Search

Free search (FS) is a very recent population-based optimizer. It was invented by K. Penev and G. Littlefair in 2003 [PL03]. Just as PSO emulates social cognition, FS is associated with an animal's behavior. FS partially imitates Ant Colony Optimization (ACO) adapted for continuous search [BP95]. Also, it includes: a PSO mechanism to refresh an animal's position, a DE strategy principle to create the animal's action and, also, the general structure of GA.

An animal in free search makes a journey, several exploration steps in the search space. Then, it moves at the best found position and marks it by a pheromone. The behavior of any animal is described by two aspects.

1. **Sense**
   Each of the animals has a sense to locate a pheromone. The more sensitive animal is able to find a better (promising) place for the search. The less sensitive one is forced to search around any marked position.
2. **Action**
   Each of the animals makes a decision of how to search; that is, it chooses its own neighborhood of search. So, the search journey of an animal may vary from local to global movements.

Both sense and action of an animal are random factors.

Let x_i be an animal of a population $\mathbb{P}$. The population consists of NP animals. The animals mark their positions by a pheromone $P_i \leq 1$:

$$P_i = f(x_i)/f_{\max} , \qquad (7.15)$$

where $f_{\max} = \max_i\{f(x_i)\}, i = 1, \ldots, NP$. Then, one endows each of the animals with the sense S_i:

$$S_i = P_{\min} + rand_i(0, 1] \cdot (P_{\max} - P_{\min}) , \qquad (7.16)$$

where $P_{\max}, P_{\min}$ are the maximum and the minimum pheromone values of a population.

At each generation the animal x_i begins its journey from any position x_k (from any animal of the population) satisfying its sense; that is,

$$x_k : \quad S_i \leq P_k, \forall k \in [1, \ldots, NP] . \qquad (7.17)$$

During the journey each animal performs T steps in the following way.

$$x_i^t = x_k + R \cdot (H - L) \cdot rand_t(0, 1), \qquad t = 1, \ldots, T . \qquad (7.18)$$

$R \in [R_{\min}, R_{\max}] \subset \mathbb{R}^D$ is a randomly generated vector that defines a neighboring space. H, L are boundary constraints. The favors of step are considered as values of the objective function $f(x_i^t)$. The animal moves itself to the best found position

$$x_i : \quad f(x_i) = \max_t\{f(x_i^t)\} .$$

When all animals perform their journeys, a new pheromone is distributed (7.15) and new senses are generated (7.16). Then, the population passes to the next generation.

This method has many random factors:

- Start position x_k
- Neighboring space R
- Steps of a journey x_i^t
- Generation of a sense S_i

In spite of the fact that the authors present such a randomness as a self-adaptation mechanism [Pen04, PL03], I suppose that it is exclusively a randomness. From my point of view self-adaptation is developed only in the restrictions on choosing a start position for a journey (7.17).

Free search can be confronted with transversal DE (Chapter 6). The introduction of a pheromone, in FS, has purely an ideological meaning. Without loss of generality the sense generation could be calculated directly from the minimal and the maximal values of an objective function. The next DE strategy will better coincide with the ideas underlying FS:

$$\omega = x + F^* \cdot (H - L), \qquad F^* = F \cdot rand(0, 1). \tag{7.19}$$

$F^* \subset \mathbb{R}^D$ is a relaxed vector of differentiation. x is a randomly extracted individual. This case presents a constant difference vector $\delta = H - L$. And, contrary to the usual DE, where extracted individuals form an exploration walk, here, only F^* moves the individual through the search space.

It is clear that there are two main disadvantages.

1. The information about the state of the population is not used; that is, the algorithm loses the perfect property of self-adaptation. In the common DE case, the difference vector is created on the basis of randomly extracted individuals that partially present the population state.
2. A random choice of the vector of differentiation produces many useless steps in the global neighborhood, and a local search is needed at the end of optimization.

However, free search introduces a new individual's feature (sense) that permits controlling the sensibility of an individual to the search space. Sense suppresses exploration of the search space, but at the same time, increases convergence by exploiting promising zones. It would be very attractive to join together the intelligence of a DE strategy and the sensitivity of a FS animal.

Problems

7.1. Formulate a definition of direct search methods. Enumerate the cases where these methods should be applied. Mention at least five direct search methods that you have already met.

7.2. What four operations of nonlinear simplex do you know? Explain and sketch in each of these operations.

7.3. Enumerate the drawbacks of the nonlinear simplex method.

7.4. Which of the strategies of differential evolution nearly completely interprets nonlinear simplex? Draw an analogy between these two algorithms.

7.5. Enumerate the advantages of differential evolution as against nonlinear simplex.

7.6. What the main idea does underlie particle swarm optimization?

7.7. Which of two appearances from psychology and sociology does particle swarm optimization reflect?

7.8. What role does the inertia weight w play in the PSO algorithm?

7.9. How is the effect of memory implemented in PSO and how is its property of forgetting implemented?

7.10. Is the memory is a positive or negative aspect of the algorithm?

7.11. Does differential evolution contain the elements of memory and personal behavior, likewise PSO? If yes, explain the difference of their implementation.

7.12. Which of the DE strategies in the best way interprets the social behaviour of PSO? Implement this strategy in your DE algorithm.

7.13. Add to problem (7.12) the implementation of the memory mechanism and the effect of personal behavior peculiar to PSO. Test the new algorithm and analyze the obtained results.

7.14. Write the algorithm that will alternate the strategy of PSO (7.13) from Chapter 7 with one of the DE strategies from Chapter 3. Experiment with alternation of one strategy with another. For the following test function,

$$f(X) = \frac{1}{2} + \frac{\left(\sin \sqrt{x_1^2 + x_2^2}\right)^2 - 1/2}{(1 + 0.001(x_1^2 + x_2^2))^2} , \quad -100 \le x_1, x_2 \le 100$$
$$f(X^*) = 0 , \quad X^* = 0 ,$$

plot the graphs of convergence depending on the percentage of one or another strategy in the algorithm. Justify the results. Compare this algorithm with the classic one from Chapter 1.

7.15. Show with an explaining sketch two main aspects of the free search algorithm.

7.16. How does one calculate the pheromone of an individual?

7.17. How does one calculate the sense of an individual?

7.18. Give a flow-graph of the free search algorithm.

7.19. What random factors does Free Search have? In your opinion are these factors advantages or drawbacks?

7.20. Which of the DE strategies coincides in the best way with the idea underlying free search? What drawbacks of this strategy do you see?

7.21. Add to your algorithm the aspect of "sense" inherent in free search. Estimate the new algorithm using, at least, the following test function

$$f(X) = (x_1^2 + x_2^2)^{0.25} \left(\sin^2 \left(50(x_1^2 + x_2^2)^{0.1} \right) + 1 \right)$$
$$-100 \le x_1, x_2 \le 100 , \quad f(X^*) = 0 , \quad X^* = 0 .$$

7.22. Find (approximate) the probability density function for the difference vector of DE. Compare this function with other familiar probability density functions. Use the obtained function to automatically generate the difference vector independently of other individuals of the population. Compare the new algorithm with the classical one.

8

Energetic Selection Principle

In this chapter[1] I shall introduce a new energetic approach and, based on it, the principle of energetic selection, which can be applied to any population-based optimization algorithm including differential evolution. It consists in both decreasing the population size and the computation efforts according to an energetic barrier function that depends on the number of generations. The value of this function acts as an energetic filter, through which can pass only individuals with lower fitness. Furthermore, this approach allows us to initialize the population of a sufficient (large) size. This method leads to an improvement of algorithm convergence.

8.1 Energetic Approach

Perhaps this new energetic approach may be associated with the processes taking place in physics. As a matter of fact, it was inspired by sociology from a certain sociobiological phenomenon, the so-called phenomenon of dispergated genes, that was observed during World War II. As only a few people know this phenomenon, I prefer to make reference to physics because it is in some sense similar and, in addition, many of the researchers working in evolutionary computation possibly know well a simulated annealing algorithm.

Let there be a population $\mathbb{P}$ consisting of NP individuals. Let us define the *potential* of an individual as its cost function value $\varphi = f(ind)$. Such a potential shows the remoteness from the optimal solution $\varphi^* = f(ind^*)$, that is, some energetic distance (potential) that should be overcome to reach the optimum. Then, the population can be characterized by superior and inferior potentials $\varphi_{\max} = \max f(ind_i)$ and $\varphi_{\min} = \min f(ind_i)$. As the population

---

[1] Some material in this chapter originally appeared in [FJ04f]; this work was selected as the best paper of ICEIS 2004 to be republished in the book *Enterprise Information Systems VI* [FJ06].

evolves the individuals take more optimal energetic positions, the closest possible to the optimum level. So if $t \to \infty$ then $\varphi_{\max}(t) \to \varphi_{\min}(t) \to \varphi^*$, where t is an elementary evolution step. Approaching the optimum, apart from stagnation cases, can also be expressed by $\varphi_{\max} \to \varphi_{\min}$ or $(\varphi_{\max} - \varphi_{\min}) \to 0$. By introducing the potential difference of population $\triangle\varphi(t) = \varphi_{\max}(t) - \varphi_{\min}(t)$ the theoretical condition of optimality is represented as

$$\triangle\varphi(t) \to 0 \,. \tag{8.1}$$

In other words, the optimum is achieved[2] when the potential difference is close to 0 or to some desired precision ε. The value $\triangle\varphi(t)$ is proportional to the algorithmic efforts, which are needed in order to find the optimal solution.

Thus, the *action* A done by the algorithm for passing from one state t_1 to another t_2 is

$$A(t_1, t_2) = \int_{t_1}^{t_2} \triangle\varphi(t)dt \,. \tag{8.2}$$

We introduce then the *potential energy* of population E_p that describes total computational expenses.

$$E_p = \int_0^\infty \triangle\varphi(t)dt \,. \tag{8.3}$$

Notice that (8.3) graphically represents the area S_p between two functions $\varphi_{\max}(t)$ and $\varphi_{\min}(t)$.

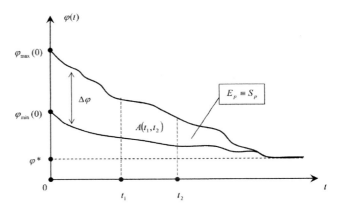

Fig. 8.1. Energetic approach.

Let us recall that our purpose is to increase the speed of algorithm convergence. Logically, convergence is proportional to computational efforts. It is obvious that the smaller the potential energy E_p is, the fewer computational efforts are needed. Thus, by decreasing the potential energy $E_p \equiv S_p$ we augment the convergence rate of the algorithm. Hence, the convergence increase is transformed into a problem of potential energy minimization (or S_p minimization).

$$E_p^* = \min_{\triangle\varphi(t)} E_p(\triangle\varphi(t)) . \tag{8.4}$$

8.2 Energetic Selection Principle

8.2.1 Idea

Now we apply the above-introduced energetic approach to the DE algorithm. As an elementary evolution step t we choose a generation g.

In order to increase the convergence rate we minimize the potential energy of population E_p (Fig. 8.1). For that a supplementary procedure is introduced at the end of each generation g. The main idea is to replace the superior potential $\varphi_{\max}(g)$ by the so-called *energetic barrier* function $\beta(g)$. Such a function artificially underestimates the potential difference of generation $\triangle\varphi(g)$.

$$\begin{aligned} &\beta(g) - \varphi_{\min}(g) \leq \varphi_{\max}(g) - \varphi_{\min}(g) \\ \Leftrightarrow\quad &\beta(g) \leq \varphi_{\max}(g), \quad \forall g \in [1, g_{\max}] . \end{aligned} \tag{8.5}$$

From an algorithmic point of view this function $\beta(g)$ serves as an *energetic filter* for the individuals passing into the next generation. Thus, only the individuals with potentials less than the current energetic barrier value can participate in the next evolutionary cycle (Fig. 8.2).

In practice, it leads to the decrease of the population size NP by rejecting individuals such that:

$$f(ind) > \beta(g) . \tag{8.6}$$

8.2.2 Energetic Barriers

Here, I shall show you some examples of the energetic barrier function. At the beginning we outline the variables upon which this function should depend. First, this is the generation variable g, which provides a passage from one evolutionary cycle to the next. Second, it should be the superior potential $\varphi_{\max}(g)$ that presents the upper bound of the barrier function. And third, it should be the inferior potential $\varphi_{\min}(g)$ giving the lower bound of the barrier function (Fig. 8.3).

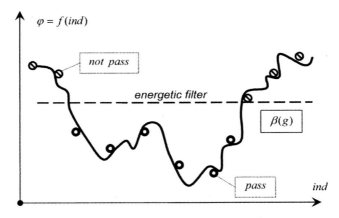

Fig. 8.2. Energetic filter.

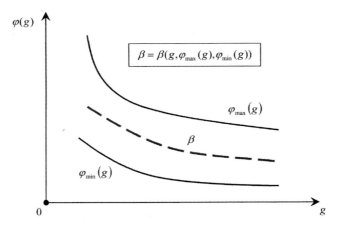

Fig. 8.3. Energetic barrier function.

Linear Energetic Barriers

The simplest example is the use of a proportional function. It is easy to obtain by multiplying either $\varphi_{\min}(g)$ or $\varphi_{\max}(g)$ with a constant K.

In the first case, the value $\varphi_{\min}(g)$ is always stored in the program as the current best value of the cost function. So, the energetic barrier looks like

$$\beta_1(g) = K \cdot \varphi_{\min}(g), \qquad K > 1. \tag{8.7}$$

The constant K is selected to satisfy the energetic barrier condition (8.5).

In the second case, a small procedure is necessary to find the superior potential (maximal cost function value of the population) $\varphi_{\max}(g)$. Here, the energetic barrier is

$$\beta_2(g) = K \cdot \varphi_{\max}(g), \qquad K < 1. \tag{8.8}$$

K should not be too small in order to provide a smooth decrease of the population size NP.

An advanced example would be a superposition of the potentials.

$$\beta_3(g) = K \cdot \varphi_{\min}(g) + (1 - K) \cdot \varphi_{\max}(g) \tag{8.9}$$

So, with $0 < K < 1$ the energetic barrier function is always found between the potential functions. Now, by adjusting K it is easier to get the smoothed reduction of the population without condition violation (8.5). Examples of the energetic barrier functions are shown in Fig. 8.4.

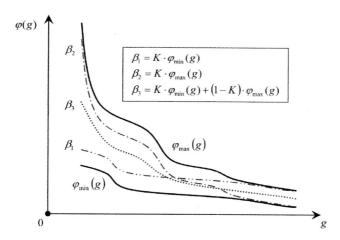

Fig. 8.4. Linear energetic barriers.

Nonlinear Energetic Barriers

As we can see, the main difficulty of using the linear barriers appears when we try to define the barrier function correctly in order to provide a desired dynamics of the population reduction. Taking into consideration that $\varphi_{\max} \rightarrow \varphi_{\min}$ when the algorithm converges locally, the ideal choice for the barrier function is a function that begins at a certain value between $\varphi_{\min}(0)$ and $\varphi_{\max}(0)$ and converges to $\varphi_{\max}(g_{\max})$.

Thereto, I propose an exponential function $K(g)$

$$K(g) = K_l + (K_h - K_l) \cdot e^{(-Tg/g_{\max})}. \tag{8.10}$$

This function, inspired by the color-temperature dependence from Bernoulli's law, smoothly converges from K_h to K_l. The constant T, so-called *temperature*, controls the convergence rate. The functional dependence on the temperature constant $K(T)$ is represented in Fig. 8.5.

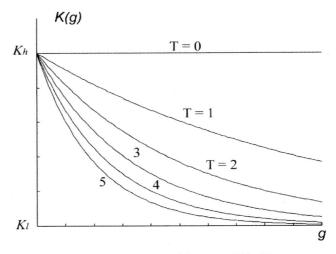

Fig. 8.5. Exponential function $K(g, T)$.

By substituting the constant K in (8.7)–(8.9) for the exponential function (8.10) we can supply the energetic barrier function with improved tuning (Fig. 8.6).

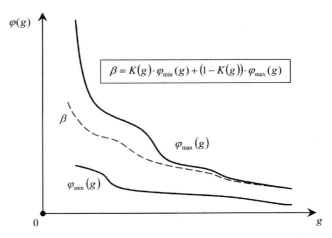

$$\beta = K(g) \cdot \varphi_{\min}(g) + (1 - K(g)) \cdot \varphi_{\max}(g)$$

Fig. 8.6. Nonlinear energetic barrier.

8.2.3 Advantages

1. The *principle of energetic selection* permits us to initialize the population of a sufficiently large size. This fact leads to better (careful) exploration

of a search space during the initial generations as well as increasing the probability of finding the global optimum.

2. The *energetic barrier function* decreases the potential energy of the population and thereby increases the convergence.
3. The *double selection principle* is applied. The first one is a usual DE selection for each individual of a population. Here, there is no reduction of the population size. And the second one is a selection of the best individuals that pass in the next generation, according to the energetic barrier function. It leads to the reduction of the population size and consequently the number of function evaluations.

Practical Remarks

Notice that a considerable reduction of the population size occurs at the beginning of the evolutionary process. For more efficient exploitation of this fact a population should be initialized with a much larger size NP_0 than usual. Then, when the population shrinks to a certain size NP_f, it is necessary to stop the energetic selection procedure. This forced stopping is explained by possible stagnation and not so efficient search by a small size population. In fact, the first group of generations locates a set of promising zones. The selected individuals are conserved in order to make a thorough local search in these zones.

8.3 Comparison of Results

In order to test this approach I took three test functions (8.11) from a standard test suite (see Appendix C). The first two functions, Sphere f_1 and Rosenbrock's function f_2, are classical DeJong testbeds [DeJ75]. The third function, rotated ellipsoid f_3, is a quadratic nonseparable function.

$$f_1(X) = \sum_{i=1}^{3} x_i^2$$
$$f_2(X) = 100(x_1^2 - x_2)^2 + (1 - x_1)^2 \qquad (8.11)$$
$$f_3(X) = \sum_{i=1}^{20} \left(\sum_{j=1}^{i} x_j \right)^2 .$$

I fixed the differentiation F and crossover Cr constants to be the same for all functions. $F = 0.5$. $Cr = 0$ (there is no crossover in order to make the DE algorithm rotationally invariant; Appendix D). The stopping condition of the algorithm is a desirable precision of optimal solution VTR (*value to reach*). It is fixed for all tests as $VTR = 10^{-6}$. As usual, we count the number of

Table 8.1. Initial test data.

f_i	D	NP	NP_0	NP_f	K
1	3	30	90	25	0.50
2	2	40	120	28	0.75
3	20	200	600	176	0.15

function evaluations NFE needed to reach VTR. The initial data are shown in Table 8.1.

For DE with the energetic selection principle the initial population size was chosen three times larger than in the classical DE scheme: $NP_0 = 3 \cdot NP$. The forced stopping was applied if the current population became smaller than NP. Hence $NP_f \leq NP$. As an energetic barrier function the linear barrier $\beta_3(g)$ was selected (8.9). So, K is an adjusting parameter for barrier tuning, which was found empirically. D is the dimension of the test functions.

·   The average results of 10 runs for both the classical DE scheme and DE with the energetic selection principle are summarized in Table 8.2.

Table 8.2. Comparison of classical differential evolution (*cl*) and differential evolution with energetic selection principle (*es*).

f_i	NFE_{cl}	NFE_{es}	$\delta, \%$
1	1088.7	912.4	16,19
2	1072.9	915.3	14,69
3	106459.8	94955.6	10,81

The numbers of function evaluations (NFEs) were compared. It is considered that the NFE_{cl} value is equal to 100%, therefore the relative convergence amelioration percentagewise can be defined as

$$\delta = 1 - \frac{NFE_{es}}{NFE_{cl}}. \tag{8.12}$$

Thus, δ may be interpreted as the improvement of the algorithm's convergence.

Remark

I tested DE with a great range of other functions. The stability of results was observed. So, in order to demonstrate my contribution, here I have generated only 10 populations for each test function relying on statistical correctness.

Problems

8.1. What is the potential of an individual? potential difference? Give an explaining sketch.

8.2. Given a test function, the so-called Schubert's problem,

$$f(X) = \prod_{i=1}^{D} \left(\sum_{j=1}^{5} j \cos((j+1)x_i + j) \right) , \quad -10 \le x_i \le 10 .$$

Plot empirical curves for both superior and inferior potentials, consider one generation (iteration) as an elementary step of evolution. Calculate the action A done by the algorithm for the first 10 and last 10 generations. Estimate the operation efficiency of the algorithm at the beginning and the ending iterations. At which moment is the algorithm most efficient? Explain why.

8.3. Calculate the potential energy of the population. As a basis for this take the curves plotted in problem (8.2).

8.4. How are the potential energy and the algorithm's convergence related?

8.5. What is the energetic barrier? Explain, how does the energetic barrier influence the population?

8.6. On what parameters does the function defining the energetic barrier depend?

8.7. Which of the linear energetic barriers do you think is most efficient from a practical point of view?

8.8. What does the constant K influence?

8.9. In which cases should you use a nonlinear energetic barrier?

8.10. What is the constant T in (8.10) of Chapter 8 and what does it influence? Using the potential's curves from problem (8.2), plot functions of nonlinear energetic barriers for the constant $T = 0, 1, 3, 5$.

8.11. Solve, for example, the following test function, the so-called McCormick's problem,

$$f(X) = \sin(x_1 + x_2) + (x_1 - x_2)^2 - 1.5x_1 + 2.5x_2 + 1$$
$$-1.5 \le x_1 \le 4 , \quad -3 \le x_2 \le 3 ,$$

using linear and nonlinear energetic barriers. Compare the obtained results.

8.12. Implement "forced stopping" of the energetic selection procedure for problem (8.11).

8.13. What are the advantages of using the method of energetic selection? For what kind of problems (test functions) is this method more appropriate? Argue your suppositions.

8.14. for determining the promising zones using the population state. Implement it in your DE algorithm.

8.15. For the algorithm realized in problem (8.14) develop a technique which permits individuals to rapidly migrate from less promising to more promising zones. Estimate the efficiency of your algorithm on 2–3 multimodal test functions at your discretion.

On Hybridization of Differential Evolution

Differential evolution is highly random in its initial form. Randomness of evolution is one of the reasons why the method is not so clever when searching for the minimum of some function $f(X)$. In this chapter I discuss how to "orient" the evolution in order to obtain a high-quality population rapidly. In particular, I propose at each generation to optimize some auxiliary function $y(X)$ that approximates the original one. This auxiliary function is constructed on the basis of the subset of "good" individuals. Many regression methods could be used to obtain $y(X)$. I have chosen least squares support vector machines (LS-SVM) because of its robustness and rapidity. The next step is to make a trade-off between classical and oriented evolution. It is clear that calculating $y(X)$ and its minimum need much more time than choosing three individuals randomly from the population. However, roughly speaking, if a random evolution needs 10^7 iterations to converge, only 10^3 iterations suffice to find nearly as good individuals with an "oriented" one. So there is hope to find a better solution within a shorter time. Besides, there is another reason that obliges us to stop the oriented evolution early. In fact, when the individuals for the construction of $y(X)$ have nearly the same values $f(X)$, the Hessian of $y(X)$ is badly conditioned. This shows that we are done with the "big" improvements and may continue with the classical evolution.

9.1 Support Vector Machine

The support vector machine (SVM) was proposed as a method of classification and nonlinear function estimation [Vap95]. The idea is to map data into higher-dimensional space, where an optimal separating hyperplane can be easily constructed. The mapping is fulfilled by means of kernel functions, which are constructed by applying Mercer's condition. In comparison with neural network methods SVMs give a global solution obtained from resolving a quadratic programming problem, whereas those techniques suffer from the existence of many local minima.

The *least squares* version of SVMs (LS-SVM) has been recently introduced [SGB+02, SV99]. There, the solution is found by solving a linear system of equations instead of quadratic programming. It results from using equality constraints in place of inequality ones. Such linear systems were named *Karush–Kuhn–Tucker* (KKT) or *augmented* systems.

Nevertheless, there remains the problem of matrix storage for large-scale tasks. In order to avoid it an iterative solution based on the conjugate gradient method has been proposed. Its computational complexity is $O(r^2)$, where r is the matrix rank. I shall mention briefly LS-SVM applied to a function approximation problem.

The LS-SVM model is represented in the feature space as

$$y(X) = \langle v, \phi(X) \rangle + b, \tag{9.1}$$

with $X \in \mathbb{R}^D$, $y \in \mathbb{R}$, and $\phi(\cdot) : \mathbb{R}^D \to \mathbb{R}^{n_h}$ is a nonlinear mapping to higher-dimensional feature space.

For given training set $\{X_k, y_k\}_{k=1}^n$ the optimization problem is formulated as

$$\min_{v,\epsilon} \Upsilon(v, \epsilon) = \frac{1}{2} \langle v, v \rangle + \gamma \frac{1}{2} \sum_{k=1}^n \epsilon_k^2 \tag{9.2}$$

subject to the *equality* constraints

$$y_k = \langle v, \phi(X_k) \rangle + b + \epsilon_k, \tag{9.3}$$

where $k = 1, \ldots, n$.

So, the Lagrangian is

$$L = \Upsilon - \sum_{k=1}^n \alpha_k \{ \langle v, \phi(X_k) \rangle + b + \epsilon_k - y_k \}, \tag{9.4}$$

where α_k are Lagrange multipliers.

By applying the Karush–Kuhn–Tucker conditions of optimality [Fle80, Fle81]

$$\frac{\partial L}{\partial v} = \frac{\partial L}{\partial b} = \frac{\partial L}{\partial \epsilon_k} = \frac{\partial L}{\partial \alpha_k} = 0, \tag{9.5}$$

the result can be transformed in matrix form

$$\begin{bmatrix} 0 & \mathbf{1}^T \\ \mathbf{1} & \Omega + \gamma^{-1} I \end{bmatrix} \begin{bmatrix} b \\ \alpha \end{bmatrix} = \begin{bmatrix} 0 \\ y \end{bmatrix}, \tag{9.6}$$

where $y = [y_1 \ldots y_n]$, $\mathbf{1} = [1 \ldots 1]$, $\alpha = [\alpha_1 \ldots \alpha_n]$, and Mercer's condition

$$\Omega_{ij} = \langle \phi(X_i), \phi(X_j) \rangle = K(X_i, X_j) \tag{9.7}$$

with $i, j = 1, \ldots, n$.

Then by solving the linear system (9.6), the approximating function (9.1) is obtained:

$$y(X) = \sum_{k=1}^{n} \alpha_k K(X_k, X) + b. \tag{9.8}$$

9.2 Hybridization

Let the objective function $f(X)$ be a nonlinear, nondifferentiable, perhaps epistatic and multimodal function. Generally, there is no winning approach to calculate the global optimum of this function by deterministic methods. Nevertheless these methods remain valid for searching local optima and provide the proof of optimality. On the other hand, metaheuristics usually give good results for the global optimum search. So, we hope to benefit from the advantages of both deterministic and metaheuristic methods.

Notice that DE is a highly random optimization method and an excessive randomness sometimes stagnates the search process. So, we would like to "orient" the evolution in order to quickly obtain a high-quality population. For this purpose it seems better to use hybridization technique. As proposed by me, hybridization incorporates one extra function without altering the structure of the DE algorithm.

The idea of hybridization consists in the deterministic creation of an individual at the end of each generation. It is intended that this individual will be near the optimum, so it replaces the worst one.

To realize such a hybridization, first a complete iteration of DE is accomplished. The obtained population better fits the surface of the objective function in the neighborhood of the optima. Then, n best individuals (*circles* in Fig. 9.1) are selected. They and their objective function values represent the training set $\{X_k, y_k\}_{k=1}^{n}$, where $y_k = f(X_k)$, for the function approximation. There exist lots of regression methods to calculate an approximation of $f(X)$. As I have already mentioned in passing, I chose the support vector machine (SVM) method [Vap95] because of its robustness inherited from regulation techniques. For recent results and implementation of SVM please see [SV99, SGB$^+$02].

The SVM regression approximation is described by (9.8), where $K(X_k, X)$ is a kernel function verifying Mercer's conditions (9.7). In order to compute α_k and b a quadratic optimization problem is resolved. The optimality (or Karush–Kuhn–Tucker) conditions of this problem are (9.6).

For a good approximation it is necessary to choose n between the dimension of individuals $(D + 2)$ and the population size NP. With a little n few computations are needed to find the support values α_k, whereas a bigger n gives better approximation quality.

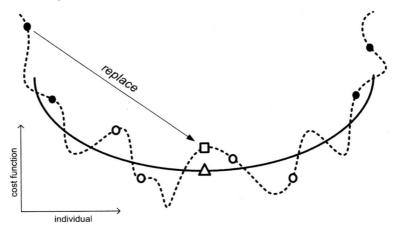

Fig. 9.1. The idea of hybridization.

The principal examples of kernels $K(X_k, X)$ are polynomials, radial basis functions, multilayer perceptrons, and others. We choose a second-order polynomial kernel function:

$$K(X_k, X) = (\langle X_k, X \rangle + 1)^2 . \tag{9.9}$$

This gives a quadratic function $y(X)$ (9.8). The optimum of the approximation (the triangle in Fig. 9.1) is thus calculated by solving a linear system. In fact, the optimality conditions are:

$$\frac{dy(X)}{dX} = \frac{\partial y(X)}{\partial X(p)} = 0 , \tag{9.10}$$

where $X(p)$, $p = 1, \ldots, D$ are components of a vector X.

These conditions (9.10) give a system of D linear equations:

$$\begin{aligned}
\frac{\partial y(X)}{\partial X(p)} &= \sum_{k=1}^{n} \alpha_k \cdot \frac{\partial K(X_k, X)}{\partial X(p)} \\
&= 2 \sum_{k=1}^{n} \alpha_k \left(\langle X_k, X \rangle + 1 \right) X_k(p) \equiv 0 \quad \Leftrightarrow \\
&\sum_{k=1}^{n} \alpha_k X_k(p) \sum_{q=1}^{D} X_k(q) X(q) \\
&= \sum_{q=1}^{D} X(q) \cdot \sum_{k=1}^{n} \alpha_k X_k(p) X_k(q) = - \sum_{k=1}^{n} \alpha_k X_k(p) .
\end{aligned} \tag{9.11}$$

Or alternatively, in matrix form

$$A \cdot X = B \,, \tag{9.12}$$

where

$$A = a(p,q) = \sum_{k=1}^{n} \alpha_k X_k(p) X_k(q)$$

$$\tag{9.13}$$

$$B = b(p) = -\sum_{k=1}^{n} \alpha_k X_k(p) \,.$$

The optimum of (9.8) is found by solving this linear system.

Notice that when individuals of the training set come too close to each other or when the gradient of $y(X)$ is nearly zero, the matrix A is badly conditioned. In this case, the results are not reliable. The obtained optimum is far from the real one. From this moment on, the matrix A is usually badly conditioned. This was numerically confirmed at the first stage of my tests (see Section 9.3). So, further application of the hybridization becomes useless and time consuming. This phenomenon serves us as a criterion for switching from the hybrid to the classical scheme (Alg. 6).

Finally, we compare the "optimum" calculated from (9.12) (the square in Fig. 9.1) with the worst individual in the population (the highest black point in Fig. 9.1). The best one enters the next generation.

This *extra function*, SVM-phase, is summarized in the following four steps.

SVM-phase:

1. The choice of n best individuals of the population, $O(n \log n)$ — sorting procedure
2. Quadratic approximation of the objective function, $O((n + 1)^3)$ — linear system of $n + 1$ equations
3. Finding the optimum of this approximation, $O(D^3)$ — linear system of D equations
4. Refreshing the population by replacing the worst individual with the obtained in Step 3 optimum, $O(NP)$ — finding the maximal element

Its overall complexity is $O(n \log n + (n + 1)^3 + D^3 + NP)$.

The hybrid algorithm consists of two phases: (1) the DE-phase and (2) the SVM-phase. The DE-phase remains as is and is performed for all iterations, whereas the SVM-phase is executed until the matrix A is badly conditioned. The pattern of the hybrid algorithm is presented below (Alg. 6).

9.3 Comparison of Results

In order to test the given approach I chose three functions from a standard test suite for evolutionary algorithms (Appendix C). The first function, rotated

Algorithm 6 Hybridization of Differential Evolution

Initialization $\mathbb{P}^0$
flag = **true**
while (not terminal condition) **do**
    DE-phase on $\mathbb{P}^g$
    if (A is well conditioned) **and** (flag) **then**
        SVM-phase on $\mathbb{P}^g$
    else
        flag = **false**
    end if
    $g \leftarrow g + 1$
end while

ellipsoid f_1, is a quadratic unimodal function. The next two functions, Rastrigin's f_2 and Ackley's f_3, are highly multimodal functions. The simulation of the algorithms is done in the MATLAB environment with the LS-SVMlab Toolbox [PSG$^+$03].

I fixed the same DE control parameters for all functions: $NP = 100$, $F = 0.5$, and $Cr = 0$. For the approximation $y(X)$ I used the whole population; that is, $n = NP$. Thus the sorting procedure to choose the best individuals is not necessary in this case.

The first time I ran the hybrid algorithm with the SVM-phase applied for each generation (Alg. 7). I used this algorithm in order to determine the critical generation η after which the matrix A generally is badly conditioned. The tests show that, for the generations $g > \eta$, the optimum of the approximation $y(X)$ rarely replaces the worst individual of the population.

Algorithm 7 Hybridization* of Differential Evolution

Initialization $\mathbb{P}^0$
while (not terminal condition) **do**
    DE-phase on $\mathbb{P}^g$
    SVM-phase on $\mathbb{P}^g$
    $g \leftarrow g + 1$
end while

The maximal number of generations $g_{\max}$ is selected here for each function separately in order to provide a good illustration of the convergence. The average results of 10 runs are summarized in Table 9.1. The convergence dynamics of these test functions is shown in Figs. 9.2–9.4 (classical algorithm — dotted line, hybrid algorithm — solid line, critical generation η — vertical line).

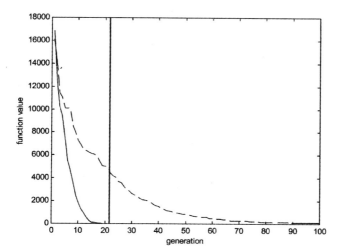

Fig. 9.2. Convergence dynamics of the rotated ellipsoid function (dotted line — classical algorithm, solid line — hybrid algorithm, vertical line — critical generation).

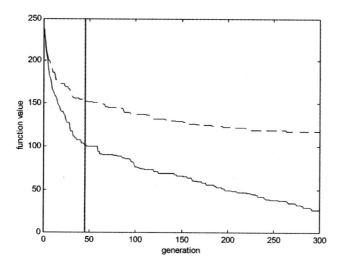

Fig. 9.3. Convergence dynamics of Rastrigin's function (dotted line — classical algorithm, solid line — hybrid algorithm, vertical line — critical generation).

Table 9.1. Comparison of both classical and hybrid approaches. D: dimension of test function; η: critical generation; DE_c^g and DE_h^g: the optimal values of the test function for classical and hybrid algorithms accordingly.

f_i	D	$g_{\max}$	η	DE_c^g	DE_h^g
1	20	100	22.1	1.070e+2	4.246e–4
2	20	300	34.8	1.170e+2	2.618e+1
3	30	100	18.8	7.713e+0	1.019e–1

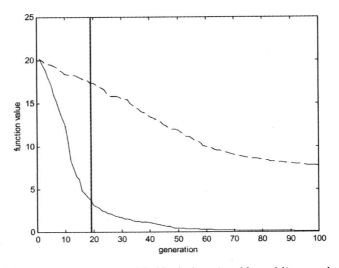

Fig. 9.4. Convergence dynamics of Ackley's function (dotted line — classical algorithm, solid line — hybrid algorithm, vertical line — critical generation).

We can see from Table 9.1 that the hybrid algorithm converges much better than the classical one. But it also seems that DE_h consumes much more time. The following tests were done with the same execution time, $t_c = t_h$, for the hybrid algorithm (Alg. 6) and the classical one. We compare the objective functions values DE_c^t and DE_h^t (Table 9.2).

As you can see from Table 9.2, the hybrid algorithm converged better. Moreover, the DE_c never reached the values of DE_h^t for the given control parameters. Thus, after these comparisons we can consider that such a hybridization is a quite promising one. Nevertheless, more tests are necessary with other functions and control parameters.

Table 9.2. The execution time is fixed; $t_c = t_h$. Comparison of the objective functions values DE_c^t and DE_h^t and of the generations g_c and g_h, accordingly.

f_i	time	DE_c^t	DE_h^t	g_c	g_h
1	10	3.151e+1	1.610e–4	366	151
2	30	8.697e+1	1.158e+1	965	602
3	20	6.776e+0	7.118e–2	565	279

9.4 Some Observations

In spite of the demonstrated potency of this approach its efficiency of approximation is not ideal. To estimate it an *efficiency measure* ζ has been introduced. ζ evaluates the percentage of successful approximation, that is, when the approximated optimum replaces the worst individual.

For the LS-SVM approach the average efficiency values are roughly 20% (see Table 9.1). In others words only 1/5 of the iterations with the SVM-phase are effective. Moreover, these are the first iterations. So, the rest of the iterations do not need such a hybridization.

As the analysis shows, this problem is caused by numerical inaccuracies while solving system (9.12). When the cost function approaches its optimum (zero in our case) this numerical effect appears. So, we are looking for other methods in order to improve the optimum approximation as well as the numerical solution of our systems.

As an example a very simple approximation could be proposed. Let us calculate, for instance, a barycenter of n best individuals.

$$X^* = \frac{1}{n} \sum_{i=1}^{n} \hat{X}_i . \tag{9.14}$$

In comparison with the LS-SVM method there are several advantages:

- No numerical peculiarities
- No inferior limits on n: $2 \leq n \leq NP$
- Very fast
- $\zeta \simeq 70\%$

Although its convergence is not as rapid as in the case of LS-SVM, for the first generations it remains faster than classical DE.

To confirm it numerically 10 runs have also been made. For each generation $n = 10$ best individuals have been chosen. The results are summarized in Table 9.3. The convergence dynamics is illustrated in Figs. 9.5–9.7 (classical algorithm — dotted line, hybrid algorithm — solid line).

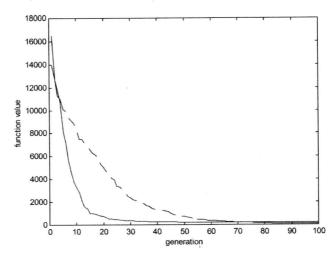

Fig. 9.5. Example of barycenter approximation: convergence dynamics of rotated ellipsoid function (dotted line — classical algorithm, solid line — hybrid algorithm).

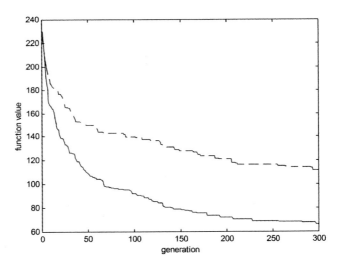

Fig. 9.6. Example of barycenter approximation: convergence dynamics of Rastrigin's function (dotted line — classical algorithm, solid line — hybrid algorithm).

Table 9.3. Comparison of the classical DE and the barycenter approximation DE_b approaches.

f_i	D	$g_{\max}$	$\zeta, \%$	DE	DE_b
1	20	100	100.	9,003e+1	2,089e+2
2	20	300	47.0	1,118e+2	6,588e+1
3	30	100	83.1	7,030e+0	8,589e+0

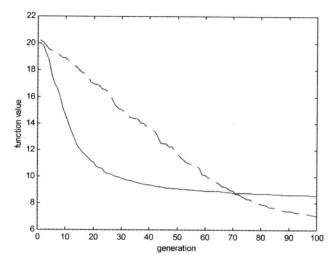

Fig. 9.7. Example of barycenter approximation: convergence dynamics of Ackley's function (dotted line — classical algorithm, solid line — hybrid algorithm).

Problems

9.1. What is the difference between interpolation and approximation? Give an example.

9.2. What methods of approximation do you know? What is the advantage of Support Vector Machines as against other methods of approximation?

9.3. For the optimization problem (9.2) and (9.3) from Chapter 9, formulate the Karush–Kuhn–Tucker conditions of optimality in matrix form. Give an explanatory sketch for this problem.

9.4. What advantages do deterministic methods of optimization have in comparison with metaheuristics?

9.5. Enumerate the main purposes of the hybridization. What means could one use to achieve these purposes. How was the hybridization described in this chapter realized?

9.6. What is the training set? What does it used for? How large should it be?

9.7. What kinds of kernel functions do you know?

9.8. Why was the second-order polynomial kernel function chosen in this chapter?

9.9. Show that the approximation obtained from (9.8) and (9.9) of Chapter 9 is the best from a practical point of view.

9.10. Propose your preferred way of approximation and find its optimal solution(s). Prove that your model is consistent.

9.11. Calculate the overall complexity of the approximation phase. Estimate whether its realization is efficient (in theory, in action)? Test on benchmarks and compare the execution time and the precision of obtained solutions.

9.12. What numerical problems can appear during execution of the algorithm? How does one overcome them?

9.13. What is the advantage of a barycenter approximation? Does it always give the best results? Explain why.

9.14. Think out your own way of a "simple" approximation. Estimate its efficiency. In which cases is it more or less preferable?

10

Applications

In order to illustrate for you the practical efficiency of differential evolution, in this chapter I present two famous problems from different fields of application. The first one is a classical identification problem in which DE is used to evaluate the parameters of a Choquet integral. This problem is a striking example from the decision-making domain. The second one belongs to the engineering design domain, the so-called bump problem. DE is applied here to find the global optimum, which is placed exactly on a constraint function. It presents the real challenge of this field. As you will see, differential evolution demonstrates prominent results on both cases in comparison with the methods that were used previously.

10.1 Decision Making with Differential Evolution

In our modern world, socioeconomical and technical, we frequently face problems having a multidimensional character. More and more we need to calculate different performances in different spaces, which can also be correlated with each other. Furthermore, we want to have a global, integral, image of our system or process often represented by an unique aggregation formula. Such a wish leads to a fuzzy representation of data that express the performances. The *Choquet integral* introduced in this section is a nice illustration of the aggregation of performances connected by dependence of subordination and of coordination. Particularly, the Choquet integral permits us to simulate the objective importance of a criterion and its interaction with the other criteria in the multicriteria space.

Here, we adapt the differential evolution algorithm to solve a classical identification problem. The task is to learn the behavior of a decision makers' group. The aggregation operator is based on the family of Choquet integrals and represents a two-additive measure. The goal is to use differential evolution as a universal algorithm that provides proximity to the optimum solution. The comparison of differential evolution with the SQP solver of MATLAB has been made. The obtained results show the following advantages.

1. The found solution is feasible.
2. It is closer to the optimum than the SQP one.

10.1.1 Aggregation by the Choquet Integral

The aggregation problem can be presented as

$$Ag : E^n \to E \quad \Leftrightarrow \quad (P_1, \dots, P_n) \to P_{Ag} = Ag(P_1, \dots, P_n), \qquad (10.1)$$

where Ag is an aggregation operator, and E^n and E are the spaces of the particular performances $(P_1, \dots, P_n)$ and their aggregation P_{Ag}, accordingly.

Before introducing some mathematical formulae I would like to emphasize the desired properties of the aggregation. I outline specially the interaction importance during aggregation. So, the next question arises: what type of interactions occur between the particular performances? It is necessary to consider that the performance indicators consist of three elements: (1) objectives, (2) measures, and (3) variables of action. The interactions could have a place on all these levels. But we shall consider only the level of objectives through which the other interactions can be induced too; that is,

$$\text{interaction (particular performances)} = \text{interaction (objectives)}. \qquad (10.2)$$

Thus, we distinguish several situations:

- The objectives are noninteractive or independent.
- The objectives are interactive.
  1. One type of interaction is a complementarity; that is, the simultaneous satisfaction of the objectives influences the aggregation performance more considerably than the separately taken satisfactions. In the other words there is a synergy.
  2. Another type of interaction is a redundancy; that is, the objectives are to some extent interchangeable. The performances of these objectives evolve to a certain degree in the same direction.

Actually, the analytical hierarchy process method (AHP) [Saa80] is widely used in practice. But unfortunately this method has some disadvantages: there is no redundancy aspect. However, the Choquet integral [Gra97] is able to simulate the described interactions.

The aggregation operators based on the family of the Choquet integral combine several operators: weighted average, min, max, medians, and ordered weighted average. This allows us to express different behaviors of the decision maker (severe, compromise, tolerant), and an importance and different interaction effects among the objectives. We consider here the particular case of the Choquet integral, the so-called two-additive measure. It represents only the pair interactions thereby simplifying the mathematical expression. Thus, the Choquet integral has the interpretable form:

$$
CI(P_1, \ldots, P_n) = \sum_{I_{ij} > 0 | i > j} \min(P_i, P_j) \cdot I_{ij}
$$

$$
+ \sum_{I_{ij} < 0 | i > j} \max(P_i, P_j) \cdot \lfloor I_{ij} \rfloor
$$

$$
+ \sum_{i=1}^{n} P_i \cdot \left(\nu_i - \frac{1}{2} \sum_{i \neq j} |I_{ij}| \right), \tag{10.3}
$$

- With $\nu_i - \frac{1}{2} \sum_{i \neq j} |I_{ij}| \geq 0$, where ν_i, $i = 1, \ldots, n$ are the Shapley indices that represent the global importance of each objective relative to the other ones; that is, $\sum_{i=1}^{n} \nu_i = 1$.
- The indices I_{ij} express the interactions between the pairs of objectives (O_i, O_j); they belong to the range $[-1, 1]$. The value 1 means that there is a positive synergistic effect between the objectives, and the value -1 means negative synergy. The value 0 signifies the independent objectives.

So, we distinguish three parts of the Choquet integral: (1) conjunctive, (2) disjunctive, and (3) additive.

1. The positive index I_{ij} results in the simultaneous satisfaction of the objectives O_i and O_j having the considerable effect on the aggregation performance and the unilateral satisfaction having no effect.
2. The negative index I_{ij} results in the satisfaction of the objectives O_i and O_j being sufficient in order to have a considerable negative effect on the aggregation performance.
3. The index I_{ij} equal to zero results in the absence of any interaction between the given objectives O_i and O_j. Also, the Shapley indices ν_i play the part of weights of the classical weighted average.

The Choquet integral has several remarkable properties. In this work we use one of them: by substituting $\max(x, y) = (x + y)/2 + |x - y|/2$ and $\min(x, y) = (x + y)/2 - |x - y|/2$ in (10.3) we deduce the next form of the Choquet integral

$$
CI(P_1, \ldots, P_n) = \sum_{i=1}^{n} P_i \cdot \nu_i - \frac{1}{2} \sum_{i > j} I_{ij} \cdot |P_i - P_j| . \tag{10.4}
$$

10.1.2 Classical Identification Problem

There are many problems that can be formulated in the context of two-additive measure aggregation. We concentrate our attention on one of them: the *classical identification problem*.

A database of couples $(u_e \in [0,1]^n, R_e \in [0,1])$ is available and the aim is to extract from this learning database the set of parameters ν_i, $i = 1, \ldots, n$ and I_{ij}, $i \neq j$, $i, j = 1, \ldots, n$ so that

$$\min_{\nu_i, I_{ij}} \sum_{e=1}^{Nl} (CI(u_e) - R_e)^2 \qquad (10.5)$$

(Nl is the cardinal of the learning set) subject to the next constraints:

$$
\begin{aligned}
&\langle 1 \rangle \quad && \sum_{i=1}^{n} \nu_i = 1 \\
&\langle 2 \rangle \quad && \nu_i - \frac{1}{2} \sum_{i \neq j} |I_{ij}| \geq 0, \quad && i = 1, \ldots, n \\
&\langle 3 \rangle \quad && \nu_i \geq 0, \quad && i = 1, \ldots, n \\
&\langle 4 \rangle \quad && |I_{ij}| \leq 1, \quad && i, j = 1, \ldots, n \\
&\langle 5 \rangle \quad && 0 \leq CI(u_e) \leq 1, \quad && e = 1, \ldots, Nl.
\end{aligned}
\qquad (10.6)
$$

To these mathematical constraints some semantic constraints may be added. They are necessarily linear inequalities such as

- $I_{ij} \leq 0$ or $I_{ij} \geq 0$ — in order to express the type of interaction between the objectives O_i and O_j.
- $I_{ij} \leq, =$ or $\geq q \cdot I_{kl}$ and $\nu_i \leq, =$ or $\geq q \cdot \nu_j$; $q \in \mathbb{N}$ — to express order of magnitude relations between the Shapley and interaction indices.

Notice the symmetry of matrix I_{ij} so that this problem has $D = n(n+1)/2$ unknowns. However, in spite of using the Choquet integral property (10.4), there is no a quadratic representation $\frac{1}{2}x^T H x + c^T x$, $Ax \leq b$ of the polynomial functional (10.5) and the constraints (10.6). Therefore, I proposed to solve this problem as a nonlinear optimization one.

10.1.3 Implementation and Comparison of Results

Now I shall explain how to solve the described identification problem by the differential evolution algorithm and then together we shall compare the obtained solutions with those given by the SQP algorithm of MATLAB.

Let us examine the constraints (10.6). We have one equality linear constraint (10.6 $\langle 1 \rangle$), $\sum_{i=1}^{n} \nu_i - 1 = 0$; n nonlinear constraints (10.6 $\langle 2 \rangle$), $\frac{1}{2} \sum_{i \neq j} |I_{ij}| - \nu_i \leq 0$; and n linear constraints (10.6 $\langle 3 \rangle$), $-\nu_i \leq 0$ and the constraints (10.6 $\langle 4 \rangle$) which can be represented as $n(n-1)$ linear constraints such as $x_k \leq 0$. And at the end, $2 \cdot Nl$ nonlinear constraints (10.6 $\langle 5 \rangle$) resulted from the Choquet integral property $0 \leq CI(u_e) \leq 1$. We don't handle the last constraints (10.6 $\langle 5 \rangle$) because the problem definition (10.5), in general, gives the solutions that do not violate these constraints. Furthermore, the semantic constraints are of no interest. They are linear and can be easily integrated to the both differential evolution and SQP methods without considerable influence on the solution. Thus, we shall handle $n(n+1)+1$ constraints. For differential evolution the part of them (10.6 $\langle 3 \rangle$ and $\langle 4 \rangle$) can be represented as boundary constraints. Taking into consideration (10.6 $\langle 1 \rangle$), ν_i belong to $[0,1]$. And the interaction indices I_{ij} lie in the range $[-1,1]$. The equality constraint we handle each time before the function evaluation in the following way: $\nu_i = \nu_i/\nu_{sum}$, $\nu_{sum} = \sum_i^n \nu_i$. So, it remains to handle explicitly only n nonlinear constraints (10.6 $\langle 2 \rangle$).

In order to test our problem we randomly generate the Nl couples (an instance) (u_e, R_e), $u_e \in [0,1]^n$ and $R_e \in [0,1]$. In this case, there are two disadvantages: (1) we don't know the global minimum of the instance and (2) we don't know about the existence of a feasible solution. However, we can accomplish the relative comparison of the both methods.

First, let us examine the exploration capabilities of differential evolution. For this, we fix the strategy $\omega = \xi_3 + F \cdot (\xi_2 - \xi_1)$, the reproduction constants $F = 0.9$ and $Cr = 0.5$, and change the size of population $NP = k \cdot D$, $k = 1.3(3), 2, 4, 6, 8, 10, 20, 40, 80, 160, 320, 640$. We run the DE algorithm 10 times in order to have the average results; each time we execute 10,000 iterations. Then, we observe the objective function values F_{obj}, the population radius R_{pop}, the execution time t, and the number of function evaluations nfe. The tests were made for $n = 5$ and $Nl = 10$. The obtained results are shown in Table 10.1.

Table 10.1. Exploration capabilities of differential evolution.

NP	F_{obj}	R_{pop}	t, msec	nfe
20	5,622e−1	9,69e−1	861	14950
30	5,619e−1	9,35e−1	1311	26822
60	5,615e−1	9,66e−1	2613	49399
90	5,611e−1	9,75e−1	3955	79030
120	**5,605e−1**	**9,88e−1**	**5267**	**107187**
150	5,610e−1	9,73e−1	6479	122475
300	5,568e−1	9,85e−1	12858	241449
600	5,611e−1	9,94e−1	25737	496313
1200	5,270e−1	1,14e+0	51594	995492
2400	4,345e−1	1,28e+0	106830	1989647
4800	*3,290e−1*	*1,42e+0*	*246610*	*3997871*
9600	4,543e−1	1,35e+0	659250	7970104

From Table 10.1 we can see how by increasing the exploration (increasing of NP) the found solution verges towards the global minimum, but the execution time becomes critical. For our further experiments we choose a compromise between the optimality of the solution and the computing time. Notice that $k = 8 \Rightarrow NP = 120$ gives a satisfactory solution for reasonable time with a good exploration potential.

Second, we vary the size of problem $n = 3, 4, 5, 6$ and compare differential evolution with the SQP method. SQP is speedier, but, as we can see below, it does not provide a feasible solution. The results are summarized in Table 10.2. Let us recall that $NP = 8 \cdot D = 4n(n + 1)$.

Table 10.2. Comparison of differential evolution with SQP (MATLAB).

n	DE F_{obj}	SQP F_{obj}	Violated Constraints
3	4,0489e−1	3,2948e−1	1–1; 2–2
4	4,9375e−1	3,3913e−1	2–4
5	1,4338e−1	8,9165e−2	2–4
6	7,5725e−1	2,0782e−1	1–1; 2–6

All DE solutions were feasible, however, all SQP solutions were infeasible. In order to show which constraints were violated we introduce the next notation $x - y$, where x is a type of constraint (10.6 $\langle x \rangle$) and y is a quantity (number) of the violated constraints. It is obvious that the main difficulty for the SQP algorithm is nonlinear constraints (10.6 $\langle 2 \rangle$), whereas differential evolution perfectly handles these constraints and always leads to a feasible solution.

Moreover, we tried to initialize the population by individuals uniformly scattered around the solution obtained by the SQP solver. We varied the scattering deviation from ± 0.2 to ± 1, but still did not find any feasible solution better than in the case of the initialization within boundary constraints. It means that the SQP solution is located far from the global optimum. Also, this fact outlines the proximity of the DE solution to the real optimum.

Summing Up

As the simulation shows, differential evolution is capable of finding a feasible solution. With increasing the search space exploration (by increasing the size of the population) the possibility of finding the global optimum rises too. The only restriction here is computing time. Moreover, as it was shown, the SQP solver finds the solution that is far from the real optimum. At the end, the method of differential evolution does not need the gradient function and uses only values of the objective function, so it is indifferent to the Choquet integral representation.

10.2 Engineering Design with Differential Evolution

10.2.1 Bump Problem

The *bump* problem is a well-known benchmark. It was first introduced by A. Keane in 1995 [Kea95]. Many research efforts have been directed towards its solving [BP96, EBK98, GHB98, Kea95, Kea96, MS96, MF02, SM96, Pen04]. The problem belongs to a class of multipeak problems that is typical for engineering design. They are hard problems. Its author's description can be found on http://www.soton.ac.uk/~ajk/bump.html.

The optimization task is written as

$$\text{maximize} \quad \frac{\left| \sum_{i=1}^{D} \cos^4(x_i) - 2 \cdot \prod_{i=1}^{D} \cos^2(x_i) \right|}{\sqrt{\sum_{i=1}^{D} i \cdot x_i^2}}$$

for

$$x_i \in [0, 10], \qquad i = 1, \ldots, D \tag{10.7}$$

subject to

$$\prod_{i=1}^{D} x_i > 0.75 \quad \text{and} \quad \sum_{i=1}^{D} x_i < 15D/2,$$

where x_i are the variables expressed in radians and D is the problem dimension. This function gives a highly bumpy surface (see Fig. 10.1 and Fig. 10.2 for $D = 2$), where the global optimum is defined by the product constraint.

10.2.2 The Best-Known Solutions

In order to compare found solutions the next limitation was accepted in [Kea95, Kea96, Pen04]:

the number of iterations is limited to $1000 \cdot D$.

The first attempts to find the solution were made by Keane [Kea95, Kea96] using a parallel genetic algorithm with 12-bit binary encoding, crossover, inversion, mutation, niche forming, and a modified Fiacco–McCormick constraint penalty function. This algorithm demonstrated the following results.

- $D = 20 \quad \Rightarrow \quad f_{max} \simeq 0.76$ after 20000 iterations.
- $D = 50 \quad \Rightarrow \quad f_{max} \simeq 0.76$ after 50000 iterations.
- More than 150,000 iterations gave $f_{max} \simeq 0.835$ for $D = 50$.

The last best-known solution was announced by K. Penev in 2004 [Pen04]. He invented a new method called free search (FS), which was inspired by

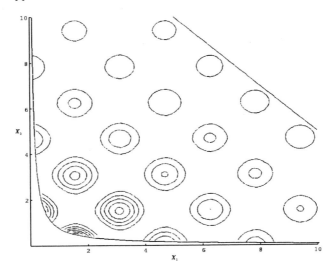

Fig. 10.1. Contour map of the two-dimensional bump problem.

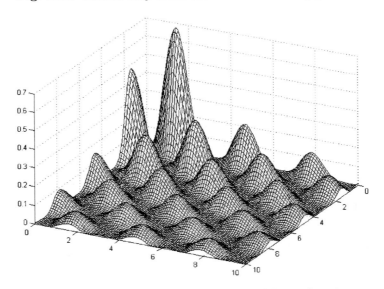

Fig. 10.2. 3-D view of the two-dimensional bump function.

Table 10.3. Free search applied to the bump problem: $D = 20$.

Initialization	Prod Constraint	Best Solution
$x_j = 5$	0.750022060021	0.80015448156
$x_j = 4 + rand_i(0,2)$	0.750000636097	0.803558343297
$x_j = 3.3 - i \cdot 0.165$	0.750015669223	0.8035972606
$x_j = x_{\max}$	0.7500013889841	0.80361832569

Table 10.4. Free search applied to the bump problem: $D = 50$.

Initialization	Prod Constraint	Best Solution
$x_j = 5$	0.750064071725	0.830462857742
$x_j = x_b$	0.750000512903	0.835238805797361
$x_j = x_{\max}$	0.750000001829	0.83526232992710514

several successful population-based metaheuristics (see Section 7.3 and [PL03, Pen04]). The achievements of this algorithm are shown in Tables 10.3 and 10.4. There, different types of initialization were used:

1. $x_j = 5$; all the individuals of a population are in the center point of the function.
2. $x_j = 4 + rand_i(0, 2)$; individuals are scattered in the center of the search space.
3. $x_j = 3.3 - i \cdot 0.165$; individuals are near the edge of the feasible region.
4. $x_j = x_b$; individuals are near the border.
5. $x_j = x_{max}$; all individuals start from the best-found solution location.

The first two types of initialization are considered as a global search, and the last three are a local search.

10.2.3 Implementation and Comparison of Results

I have applied differential evolution to solve this problem. The random initialization within the boundary constraints $x_i \in [0, 10]$ has been used throughout. It could be considered as an equivalent of the initialization by the center point in FS, because at each FS iteration the individuals execute a random walk within the boundary constraints.

As for the constraints, I used the multiobjective approach stated in Subsection 2.6.2. The product constraint $\prod_{i=1}^{D} x_i > 0.75$ is the principal one that we need to handle. The sum constraint $\sum_{i=1}^{D} x_i < 15D/2$ usually is always satisfied.

First, I tried the classical strategy with a typical set of control parameters. For $D = 20$ it resulted in

the objective function $f_{\max} = 0.80361910412556048$,
the product constraint $p_c = 0.75000000000004041$.

That is already better than the last declared solution (Table 10.3). Further research led me to the following improvements.

- Relaxation on the differentiation constant, $F^* = F \cdot rand(0, 1]$
- Appropriate choice of the strategies
  RAND group for a global search and
  RAND/BEST group for a local search

- Alternation of the strategies during the search
- Several restarts with different values of control parameters

All this allowed us to find the next optimal points.

Table 10.5. The best solutions found by DE.

Dim	Solution	Prod Constraint	Sum Constraint
20	0.80361910412558857	0.7500000000000022	29.932583972573589
50	0.83526234835804869	0.7500000000000500	78.000115253998743

Table 10.6. The vector of optimal parameters for $D = 20$.

$x_{00} = 3.16246065811604190$	$x_{01} = 3.12833145017448370$
$x_{02} = 3.09479214084494900$	$x_{03} = 3.06145059811428410$
$x_{04} = 3.02792919854217060$	$x_{05} = 2.99382609801282040$
$x_{06} = 2.95866872595774040$	$x_{07} = 2.92184226455266010$
$x_{08} = 0.49482513593227867$	$x_{09} = 0.48835709655412185$
$x_{10} = 0.48231641293878313$	$x_{11} = 0.47664472480676473$
$x_{12} = 0.47129549301840445$	$x_{13} = 0.46623099767787296$
$x_{14} = 0.46142006012626491$	$x_{15} = 0.45683663647966960$
$x_{16} = 0.45245879070189926$	$x_{17} = 0.44826762029086792$
$x_{18} = 0.44424701500322117$	$x_{19} = 0.44038285472829569$

I verified these solutions by a gradient method (MATLAB). Starting from these points the local searcher could not find a better solution. In fact, 64-bit encoding provides numerical validity for tolerance more than or equal to 1.0e–16. So, my solutions can be considered as the global optima.

At the end, it is reasonable to compare the obtained solutions with the best-known ones (Tables 10.3 and 10.4). Logically, they should be compared with the solutions obtained from the center point initialization (type 1); it ensures nearly the same start conditions and demonstrates the capabilities of a global exploration of the algorithms. The comparative results are summarized in Table 10.8. Obviously the one proposed by my DE algorithm outperforms the last winner. Moreover, it attains better performance than FS initialized[1] by the previously obtained best solution.

---

[1] Initialization:

Center — individuals start from a center point of the function, $x_j = 5$.

Best — individuals start from the previously found best solution, $x_j = x_{\max}$.

Global — initialization within boundary constraints, $x_j = L + rand_j[L, H]$.

Table 10.7. The vector of optimal parameters for $D = 50$.

$x_{00} = 6.28357974793778330$	$x_{01} = 3.16993733816407190$
$x_{02} = 3.15607465250342400$	$x_{03} = 3.14236079172872750$
$x_{04} = 3.12876948312050820$	$x_{05} = 3.11527494621216540$
$x_{06} = 3.10185302117381670$	$x_{07} = 3.08848016178066940$
$x_{08} = 3.07513491216427640$	$x_{09} = 3.06179467252986190$
$x_{10} = 3.04843675871158260$	$x_{11} = 3.03503848140045960$
$x_{12} = 3.02157775558343110$	$x_{13} = 3.00802924772837340$
$x_{14} = 2.99436736737406540$	$x_{15} = 2.98056473717553460$
$x_{16} = 2.96659073290957930$	$x_{17} = 2.95241155210926020$
$x_{18} = 2.93799064523733120$	$x_{19} = 2.92328402602021910$
$x_{20} = 0.48823764243053802$	$x_{21} = 0.48593348169600903$
$x_{22} = 0.48368276001618171$	$x_{23} = 0.48148238865686760$
$x_{24} = 0.47932961305325728$	$x_{25} = 0.47722233714893758$
$x_{26} = 0.47515884532321256$	$x_{27} = 0.47313717927816629$
$x_{28} = 0.47115523634530548$	$x_{29} = 0.46921225802399896$
$x_{30} = 0.46730549186047610$	$x_{31} = 0.46543422118642180$
$x_{32} = 0.46359702444691414$	$x_{33} = 0.46179207306529080$
$x_{34} = 0.46001901985326699$	$x_{35} = 0.45827617833553030$
$x_{36} = 0.45656228870835214$	$x_{37} = 0.45487673783362242$
$x_{38} = 0.45321838205667508$	$x_{39} = 0.45158652641468644$
$x_{40} = 0.44997988261943855$	$x_{41} = 0.44839838037402802$
$x_{42} = 0.44684083652996631$	$x_{43} = 0.44530577084165557$
$x_{44} = 0.44379357586475304$	$x_{45} = 0.44230327040711964$
$x_{46} = 0.44083416284896781$	$x_{47} = 0.43938573760010979$
$x_{48} = 0.43795618165024869$	$x_{49} = 0.43654673796357019$

Table 10.8. Comparison of DE with FS.

Dim	Algorithm	Initialization	Solution
20	FS	Center	0.80015448156000000
		Best	0.80361832569000000
	DE	Global	0.80361910412558857
50	FS	Center	0.83046285774200000
		Best	0.83526232992710514
	DE	Global	0.83526234835804869

11

End Notes

For the last ten years, the methods of metaheuristic optimization have been enjoyed wide popularity and recognition in the worlds of science, business, and industry. In particular, evolutionary algorithms seem to impose themselves as the better choice method for optimization problems that are too intricate to be solved by traditional techniques. They are universal, robust, easy to use, and intrinsically parallel. Nobody can ignore the great number of applications and the permanent interest devoted to them. In practice, a very large number of these methods are inspired by nature, physics, or psychology. The problem of the choice of method and of evaluation of its performance is becoming a problem in and of itself. Who knows, a priori, if this method is better than another for a particular problem? Outwardly, algorithms are often very similar and although, at the same time, so different when the analysis becomes finer. Can we reveal the principal patterns that unify them? Can we indicate the tendencies of their evolution? Can we see prospectively and predict the horizons of their future development?

This book seems to partially answer these questions. After a profound analysis of the optimization domain, my attention was attracted by an algorithm, universally recognized at that time and quite famous now, named differential evolution. This is a method of population-based optimization. It synthesizes today the state of the art of evolutionary computation. I analyzed this algorithm deeply and improved its performance (convergence and precision). The present studies introduce a certain number of elements helping to comprehend and answer some previously posed questions.

- Discovery of the origins of the success of the algorithm (operation differentiation) that led to:

  1. Elaboration of new strategies, their classification, and generalization by a unique formula (Chapter 3). Now, differential evolution is no longer reduced to a certain set of strategies. Strategies can be created depending on the properties of an optimization task. The introduced

classes will direct the user to a better choice of the type of strategy (random/directed/local/hybrid). In fact, there is no longer a need to hesitate between differential evolution and other population-based optimizers. DE strategies, namely the flexibility in their creation, could, ideally, imitate other well-known methods of optimization (Chapter 7). Moreover, most comparative tests have shown the numerical superiority of differential evolution.

2. Analysis of the trade-off between exploration and exploitation capacities of the algorithm and, in particular, of differentiation. I have introduced a probabilistic measure of the strategy's diversity (Chapter 4) and proved numerically that the diversity of a population should decrease proportionally to approaching the optimum (Chapter 6). Also, I propose that differentiation is the first step towards the general operator integrating the features of both mutation and crossover, for which the evolutionary computation community has been looking for a long time (Section 4.1). Principal methods of diversity evaluation, adaptation of control parameters, and convergence improvement have been considered as well.

3. Introduction of new measures to evaluate more objectively the behavior of the algorithm (or a strategy) (Chapter 5). Three new measures have been proposed. (1) *Q-measure:* an integral criterion to measure the convergence of an objective function. (2) *R-measure:* a statistical criterion to measure the robustness of a strategy with respect to control parameters. (3) *P-measure:* a dynamic criterion to measure the radius of a population, which then characterizes the population convergence.

- Discovery of three levels of performance improvement of the algorithm, namely:

  1. *Individual Level*

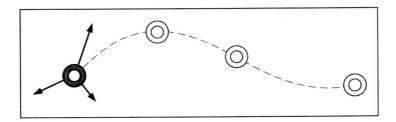

On this level, I have proposed a new vision of differentiation (see above) and introduced a new DE species, named transversal differential evolution (Chapter 6). There are three reasons to introduce the transverse species:

(a) *Diversity control:* varying the number of transversal steps.

(b) *Universalization:* a transversal architecture can correctly imitate some other algorithms, for example, recently invented free search; on the other hand, three denoted species (two-array, sequential, and transversal) enlarge the concept of differential evolution; for instance, the sequential species is now a particular case of the transversal one.

(c) *Efficient parallelization:* a transversal architecture procures much more flexibility and quality in comparison with others for implementation on heterogeneous networks of computers.

2. *Population Level*[1]

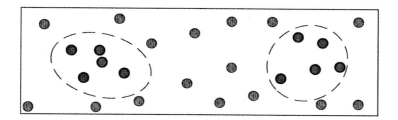

On this level, the principle of energetic vitality of individuals is developed (Chapter 8). At each generation, the population passes an energetic obstacle that rejects feeble individuals. The proposed energetic approach explains the theoretical aspect of such a population size reduction. The present innovation provides more thorough exploration of a search space and accelerates convergence, which, in its turn, augments the probability of finding the global solution.

3. *External Level*

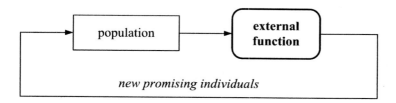

This level embodies an interaction of differential evolution with other external techniques. Undoubtedly, hybridization of differential evolu-

---

[1] Note: The innovations on population and external levels can be applied to any population-based algorithm.

tion with regression methods is a rather promising tendency. The principal idea consists in finding a better solution on the basis of potentially good individuals. The obtained solution will replace the worst. Such an iterative refreshment of a population leads to increasing the algorithm's convergence. I have tested a combination of differential evolution with support vector machines (Chapter 9). The introduction of a polynomial second-order kernel permitted the calculation of the global optimum of an approximation function by solving a linear system of equations. Also, for the sole purpose of comparison, I have used a simple barycentric approximation. In both cases, increase of the convergence has been discovered. The key elements that influence convergence are (1) an appropriate choice of a kernel function and (2) a choice of an equation solver for the linear system.

- Comparison with some other well-known algorithms. Such a comparison permitted clearly underlining advantages and disadvantages of differential evolution and revealing common and distinguishing points with other algorithms. In the book, as a case in point, I have interpreted three of the best known algorithms (nonlinear simplex, particle swarm optimization, and free search) through differential evolution (Chapter 7).

Below you will find my principal contribution to differential evolution in brief.

1. Introduction of the universal formula of differentiation
2. Classification of the strategies (random/directed/local/hybrid)
3. Uncovering of the transversal DE species
4. Universalization of the algorithm
5. Development of the energetic approach (energetic selection)
6. Hybridization differential evolution with regression methods (SVM)
7. Suggestion of new algorithm performance measures (Q-, R-, P-measures)
8. Analysis and generalization of some other methods through DE
9. Applications (decision making, engineering design)

In this book I expounded the material as completely as possible. However, if after reading it you have some questions, please do not hesitate to contact me by e-mail, Vitaliy.Feoktistov@gmail.com. I will soon be creating a Web site and shall put my algorithms there for free use. I hope sincerely that this book will help you in your work. Read and achieve success!

A

Famous Differential Evolution

A.1 C Source Code

The code is partitioned for convenience on three files. The first file "main.c" is a main program, and the other two files "rand.h" and "function.h" are header files. In "rand.h" you can place your favorite random number generator. And in "function.h" you can put any function for testing. Of course, this code is rather simplified for study purposes and is rich in comments. It clearly reflects the basic principles of differential evolution.

main.c

```c
#include <stdio.h>
#include "rand.h"          // random number generator
#include "function.h"      // objective function (fitness)

// ** CONTROL PARAMETERS ** //
#define D    10            // dimension of problem
#define NP   60            // size of population
#define F    0.9           // differentiation constant
#define CR   0.5           // crossover constant
#define GEN  10000         // number of generations
#define L    -2.048        // low boundary constraint
#define H    2.048         // high boundary constraint


int main() {

//***************************//
//** ALGORITHM'S VARIABLES **//
//***************************//
```

```
double X[D] ;           // trial vector
double Pop[D][NP] ;     // population
double Fit[NP] ;        // fitness of the population
double f ;              // fitness of the trial individual
int iBest = 0 ;         // index of the best solution
int i,j,g ;             // loop variables
int Rnd ;               // mutation parameter
int r[3] ;              // randomly selected indices

//****************************//
//** CREATION OF POPULATION **//
//****************************//

ini_rand(654987654UL) ; // initialize rand
for (j=0; j<NP; j++)    // initialize each individual
{
    for (i=0; i<D; i++) // within boundary constraints
        Pop[i][j] = X [i] = L + (H-L)*rand() ;
    Fit[j] = fnc(D,X) ; // and evaluate fitness function
}

//*****************//
//** OPTIMIZATION **//
//*****************//

for (g=0; g<GEN; g++)      // for each generation
{
    for (j=0; j<NP; j++)   // for each individual
    {
        // choose three random individuals from population,
          // mutually different and also different from j
        r[0] = (int) (rand()*NP) ;
        while (r[0]==j)
                r[0] = (int) (rand()*NP) ;
        r[1] = (int) rand()*NP ;
        while ((r[1]==r[0])||(r[1]==j))
                r[1] = (int) (rand()*NP) ;
        r[2] = (int) (rand()*NP) ;
        while ((r[2]==r[1])||(r[2]==r[0])||(r[2]==j))
                r[2] = (int) (rand()*NP) ;

        // create trial individual
          // in which at least one parameter is changed
        Rnd = (int)(rand()*D) ;
```

```
          for (i=0; i<D; i++)
          {
              if ( (rand()<CR) || (Rnd == i) )
                  X[i] = Pop[i][r[2]] +
                      F * (Pop[i][r[0]] - Pop[i][r[1]]) ;
              else
                  X[i] = Pop[i][j] ;
          }

          // verify boundary constraints
          for (i=0; i<D; i++)
              if ((X[i]<L)||(X[i]>H))
                  X[i] = L + (H-L)*rand() ;

          // select the best individual
          // between trial and current ones
            // evaluate fitness of trial individual
          f = fnc(D,X) ;
            // if trial is better or equal than current
          if (f <= Fit[j])
          {
                // replace current by trial
              for (i=0; i<D; i++)
                  Pop[i][j] = X[i] ;
              Fit[j] = f ;

                // if trial is better than the best
              if (f <= Fit[iBest])
                  iBest = j ;  // update the best's index
          }
      }
  }
}

//*************//
//** RESULTS **//
//*************//

printf("OPTIMUM : \n");
for (i=0; i<D; i++)
    printf("%g\n",Pop[i][iBest]);
printf("Fobj = %g\n",Fit[iBest]);

scanf("%hd",&i);
return 0;
```

rand.h

```
// Period parameters //
#define N 624
#define M 397
#define MATRIX_A 0x9908b0dfUL        // constant vector a //
#define UMASK 0x80000000UL // most significant w-r bits //
#define LMASK 0x7fffffffUL  // least significant r bits //
#define MIXBITS(u,v) ( ((u) & UMASK) | ((v) & LMASK) )
#define TWIST(u,v)
        ((MIXBITS(u,v) >> 1) ^ ((v)&1UL ? MATRIX_A : 0UL))

static unsigned long state[N]; // state vector  //
static int left = 1;
static int initf = 0;
static unsigned long *next;

/* initializes state[N] with a seed */
void ini_rand(unsigned long s)
{
    int j;
    state[0]= s & 0xffffffffUL;
    for (j=1; j<N; j++) {
        state[j] = (1812433253UL *
                (state[j-1] ^ (state[j-1] >> 30)) + j);
    /* See Knuth TAOCP Vol2. 3rd Ed. P.106 for multiplier. */
    /* In the previous versions, MSBs of the seed affect   */
    /* only MSBs of the array state[].                      */
    /* 2002/01/09 modified by Makoto Matsumoto              */
        state[j] &= 0xffffffffUL;  /* for >32 bit machines */
    }
    left = 1; initf = 1;
}

static void next_state(void)
{
    unsigned long *p=state;
    int j;

    // if ini_rand() has not been called, //
    // a default initial seed is used     //
    if (initf==0) ini_rand(5489UL);

    left = N;
    next = state;
```

```
    for (j=N-M+1; --j; p++)
        *p = p[M] ^ TWIST(p[0], p[1]);

    for (j=M; --j; p++)
        *p = p[M-N] ^ TWIST(p[0], p[1]);

    *p = p[M-N] ^ TWIST(p[0], state[0]);
}


/* generates a random number on [0,1)-real-interval */
double rand(void)
{
    unsigned long y;

    if (--left == 0) next_state();
    y = *next++;

    /* Tempering */
    y ^= (y >> 11);
    y ^= (y << 7) & 0x9d2c5680UL;
    y ^= (y << 15) & 0xefc60000UL;
    y ^= (y >> 18);

    return (double)y * (1.0/4294967296.0);
    /* divided by 2^32 */
}
```

function.h

```
    double fnc(int D, double* X)
    {
            double f = 0;
            int i;
            for (i=0; i<D-1; i++)
                f += 100*(X[i]*X[i]-X[i+1])*(X[i]*X[i]-X[i+1])
                    + (1-X[i])*(1-X[i]);
            return f;
    }
```

A.2 MATLAB Source Code

Recently MATLAB has become more and more popular among researchers, engineers, and students; I decided to include MATLAB code also.

```
function [f,X] = DE
    % f - optimal fitness
    % X - optimal solution

% CONTROL PARAMETERS %
D   = 10;              % dimension of problem
NP  = 60;              % size of population
F   = 0.9;             % differentiation constant
CR  = 0.5;             % crossover constant
GEN = 10000;           % number of generations
L   = -2.048;          % low boundary constraint
H   = 2.048;           % high boundary constraint

% ************************** %
% ** ALGORITHM'S VARIABLES ** %
% ************************** %

X = zeros(D,1);      % trial vector
Pop = zeros(D,NP);   % population
Fit = zeros(1,NP);   % fitness of the population
iBest = 1;           % index of the best solution
r = zeros(3,1);      % randomly selected indices

% ********************** %
% ** CREATE POPULATION ** %
% ********************** %

% initialize random number generator
rand('state',sum(100*clock));
for j = 1:NP                    % initialize each individual
    Pop(:,j) = L + (H-L)*rand(D,1); % within b.constraints
    Fit(1,j) = fnc(Pop(:,j));        % and evaluate fitness
end

% **************** %
% ** OPTIMIZATION ** %
% **************** %

for g = 1:GEN        % for each generation

    for j = 1:NP     % for each individual

        % choose three random individuals from population,
            % mutually different and different from j
```

```
r(1) = floor(rand()* NP) + 1;
while r(1)==j
   r(1) = floor(rand()* NP) + 1;
end
r(2) = floor(rand()* NP) + 1;
while (r(2)==r(1))||(r(2)==j)
    r(2) = floor(rand()* NP) + 1;
end
r(3) = floor(rand()* NP) + 1;
while (r(3)==r(2))||(r(3)==r(1))||(r(3)==j)
    r(3) = floor(rand()* NP) + 1;
end

% create trial individual
% in which at least one parameter is changed
   Rnd = floor(rand()*D) + 1;
 for i = 1:D
     if ( rand()<CR ) || ( Rnd==i )
        X(i) = Pop(i,r(3)) +
           F * (Pop(i,r(1)) - Pop(i,r(2)));
     else
         X(i) = Pop(i,j);
     end
 end

% verify boundary constraints
for i = 1:D
    if (X(i)<L)||(X(i)>H)
        X(i) = L + (H-L)*rand();
    end
end

% select the best individual
% between trial and current ones
   % calculate fitness of trial individual
f = fnc(X);
   % if trial is better or equal than current
if f <= Fit(j)
  Pop(:,j) = X;         % replace current by trial
  Fit(j) = f ;
      % if trial is better than the best
  if f <= Fit(iBest)
        iBest = j ;    % update the best's index
  end
end
```

```
        end
end

% ************* %
% ** RESULTS ** %
% ************* %

f = Fit(iBest);
X = Pop(:,iBest);


% ============================================= %
function f = fnc(X)
    % fitness function
n = length(X);
f = 0;
for i = 1:n-1
    f = f + 100 * ( X(i,1)*X(i,1) - X(i+1,1))^2
        + (X(i,1) - 1)^2;
end
```

B

Intelligent Selection Rules

This is a source code written in C language of the modified selection operation, selection rules, explained in Section 2.6. The function "*selection*" returns *true* if the trial individual is chosen, otherwise it returns *false*. The entry function argument is a population index j (see Alg. 1) of the current individual.

```c
bool selection(int currentInd)
{

    bool trialFeasible = true;
    bool currentFeasible = true;
    bool choiceTrial = true;

    // constraints verification
    for (int i=0; i!=NCONSTR; i++)
    {
        // evaluate i-th constraint
        constr[i] = fconstraints(i);

        // if the i-th constraint is dominant?
        if ((constr[i]>=0.0) &&
                    (constr[i]>cpop[i][currentInd]))
            return choiceTrial = false;

        // feasibility of trial and current solutions
        if (constr[i] >= 0.0)
                            trialFeasible = false;
        if (cpop[i][currentInd] >= 0.0)
                            currentFeasible = false;
    }
```

```
    // evaluate the objective function
    fi = fobj(ind);

    // if both solutions are infeasible => trial choice
    if ((!trialFeasible)&&(!currentFeasible))
        return choiceTrial;

    // if the both solutions are feasible =>
        // comparison of the fobj values
    if ((currentFeasible)&&(fi<fpop[currentInd]))
        return choiceTrial = false;

    return choiceTrial;
}
```

C

Standard Test Suite

Here, in this appendix, a standard test suite is presented[1]. The illustrated test functions were used to compare proposed strategies and to validate the other ideas discussed in this monograph. The first five test functions $f_1 - f_5$ (Figs. C.1–C.5) were introduced more than 20 years ago by DeJong [DeJ75]. These are classical examples representing different kinds of difficulties for an evolutionary solver. The next two functions f_6, f_7 (Figs. C.6 and C.7) are examples of a highly multimodal search space. They contain millions of local optima in the interval of consideration. The final f_8 function (Fig. C.8) is a difficult quadratic optimization problem.

C.1 Sphere Function

The function sphere, f_1, (Fig. C.1) is the "dream" of every optimization algorithm. It is a smooth, unimodal, and symmetric function and it does not present any of the difficulties that we have discussed so far. The performance on the sphere function is a measure of the general efficiency of an algorithm.

$$f_1(X) = \sum_{i=1}^{n} x_i^2 , \qquad -5.12 \leq x_i \leq 5.12 . \qquad (C.1)$$

C.2 Rosenbrock's Function

The second one, Rosenbrock's function (Fig. C.2) is very bad. It has a very narrow ridge. The tip of the ridge is very sharp, and it runs around a parabola. The algorithms that are not able to discover good directions underperform in this problem.

---

[1] You can find more test problems in [FPA+99].

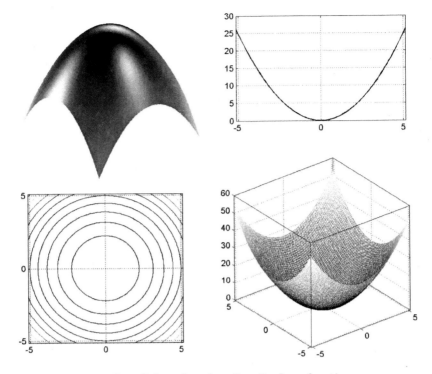

Fig. C.1. Sphere function, first De Jong function.

$$f_2(X) = \sum_{i=1}^{n-1} 100 \cdot (x_i^2 - x_{i+1})^2 + (1 - x_i)^2, \qquad -2.048 \le x_i \le 2.048. \quad \text{(C.2)}$$

C.3 Step Function

The third, the step function (Fig. C.3), is representative of the problem of flat surfaces. Flat surfaces are obstacles for optimization algorithms because they do not give any information as to which direction is favorable. The derivations are zero. Unless an algorithm has variable step sizes, it can get stuck on one of the flat plateaus.

$$f_3(X) = \sum_{i=1}^{n} \lfloor x_i \rfloor, \qquad -5.12 \le x_i \le 5.12. \quad \text{(C.3)}$$

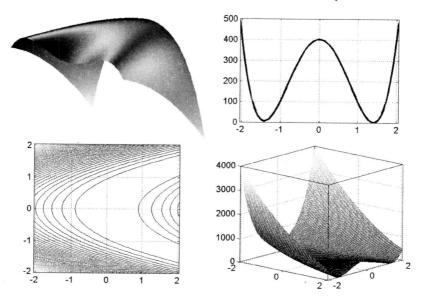

Fig. C.2. Rosenbrock's function, second De Jong function.

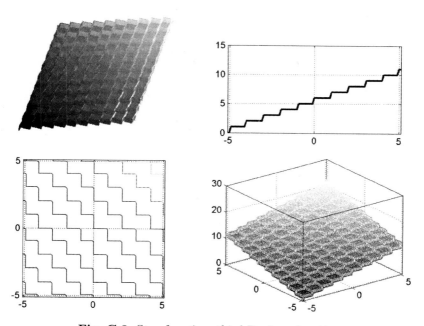

Fig. C.3. Step function, third De Jong function.

C.4 Quartic Function

The fourth is the quartic function (Fig. C.4). This is a simple unimodal function padded with noise. The Gaussian noise makes sure that the algorithm never gets the same value on the same point. The algorithms that do not perform well on this test will work poorly on noisy data.

$$f_4(X) = \sum_{i=1}^{n} \left(i \cdot x_i^4 + Gauss(0, 1) \right) , \qquad -1.28 \leq x_i \leq 1.28 . \tag{C.4}$$

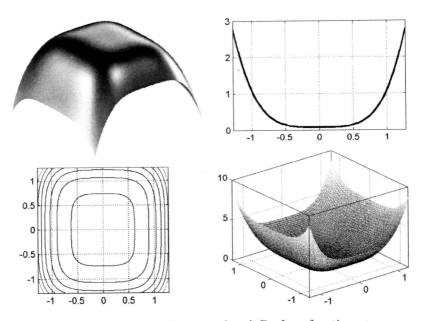

Fig. C.4. Quartic function, fourth De Jong function.

C.5 Shekel's Function

The fifth function (Fig. C.5), is named Shekel's function or foxholes. It is an example of many (in this case 25) local optima. Many optimization algorithms get stuck in the first peak they find.

$$f_5(X) = \frac{1}{0.002 + \sum_{j=1}^{25} \frac{1}{c_j + \sum_{i=1}^{2}(x_i - a_{ij})^6}} , \qquad -65.536 \leq x_i \leq 65.536 . \tag{C.5}$$

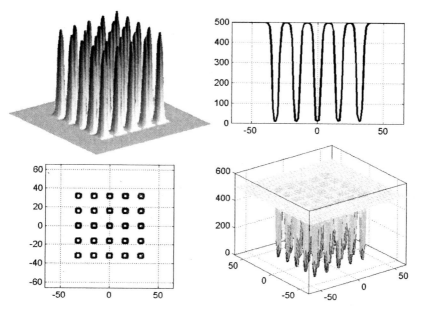

Fig. C.5. Shekel's function (foxholes), fifth De Jong function.

C.6 Rastrigin's Function

This function, the so-called Rastrigin's function (Fig. C.6), is an example of a highly multimodal search space. It has several hundred local optima in the interval of consideration.

$$f_6(X) = \sum_{i=1}^{n} \left[x_i^2 - 10\cos(2\pi x_i) + 10 \right] , \qquad -5.12 \leq x_i \leq 5.12 . \qquad (C.6)$$

C.7 Ackley's Function

This is the next highly multimodal function, Ackley's function (Fig. C.7).

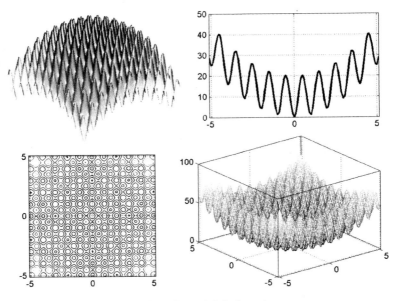

Fig. C.6. Rastrigin's function.

$$f_7(X) = -20 \exp\left(-0.2\sqrt{\frac{1}{n}\sum_{i=1}^{n}x_i^2}\right) -$$

$$- \exp\left(\frac{1}{n}\sum_{i=1}^{n}\cos(2\pi x_i)\right) + 20 + \exp, \qquad \text{(C.7)}$$

$$-32.768 \leq x_i \leq 32.768$$

C.8 Rotated Ellipsoid Function

$$f_8(X) = \sum_{i=1}^{n}\left(\sum_{j=1}^{i}x_i\right)^2, \qquad -65.536 \leq x_i \leq 65.536. \qquad \text{(C.8)}$$

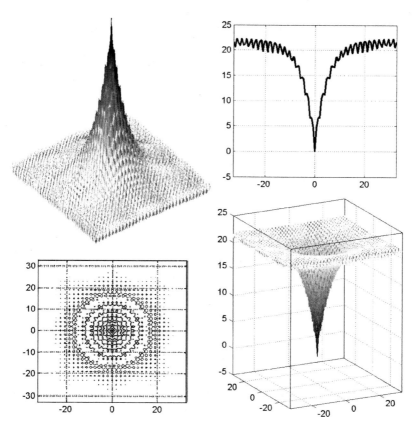

Fig. C.7. Ackley's function.

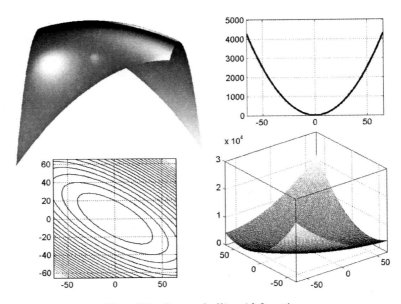

Fig. C.8. Rotated ellipsoid function.

D

Coordinate Rotation of a Test Function

An efficient technique to struggle against the disadvantages of test functions is presented here. The matter concerns a rotation of the function's coordinates proposed by Salomon [Sal96]. In particular, it permits us to transform a separable function into a highly epistatic one.

Most of the test functions (Appendix C) are separable in the following sense,

$$f(X) = \sum_{i=1}^{D} f_i(x_i) \, . \tag{D.1}$$

If a function is separable, then its variables are independent. Such a function has the following property

$$\forall i, j \quad i \neq j : \quad f(\ldots, x_i^o, \ldots) = opt \quad \wedge \quad f(\ldots, x_j^o, \ldots) = opt \\ \Rightarrow \quad f(\ldots, x_i^o, \ldots, x_j^o, \ldots) = opt \, . \tag{D.2}$$

This property makes it easy to optimize a separable function. The optimization could be done in a sequence of D independent processes resulting in D independent optimal values. Then the solution is the combination of these values. That gives the complexity equal to $O(D)$.

The functions, such as $f(X) = (x_1^2 + x_2^2)^2$, are not separable. However, these functions are also easy to optimize because their first derivative is a product, such as $\partial f(X)/\partial x_1 = 4x_1(x_1^2 + x_2^2)$. Such a product gives a solution $x_1 = 0$ that is independent of the other variable. In general, all functions that fulfill the following condition,

$$\frac{\partial f(X)}{\partial x_i} = g(x_i) \cdot h(X) \, , \tag{D.3}$$

are as easy to optimize as separable functions, because they permit us to obtain solutions for each x_i independently of all other variables. The property (D.1) is a special case of (D.3).

All functions satisfying condition (D.3) can be solved with an $O(n)$ complexity. Salomon showed that many GAs that use a small mutation probability $p_m \leq 1/D$ have the complexity $O(n \ln n)$ [Sal96]. That is greater than $O(n)$, but seems to be optimal for a randomized search.

This fact can be explained in the following way. Independent variables permit us to decompose a D-dimensional task into a sequence of D independent one-dimensional ones. This set of one-dimensional tasks can be scheduled in any arbitrary sequence. A GA with a small mutation probability does this scheduling by modifying only one variable at a time. Such a mechanism works with an $O(n \ln n)$ complexity as long as the variables are independent, but suffers a drastic performance loss as soon as the variables become dependent on each other.

Figure D.1 explains the performance loss. The left part shows a quadratic function of the form $f(X) = x_1^2 + \alpha_1 x_2^2$ that is aligned with the coordinate system. The right part shows the same function but rotated, which leads to $f(X) = x_1^2 + \alpha_0 x_1 x_2 + \alpha_1 x_2^2$. It can be clearly seen that the improvement intervals are rapidly shrinking, which results in an increase of the algorithm's complexity. This is a simple two-dimensional example with a very small eccentricity. The observable performance loss is much higher in high-dimensional search space and with large eccentricities. In summary, the efficiency of the algorithm is sensitive to the axis rotation angle.

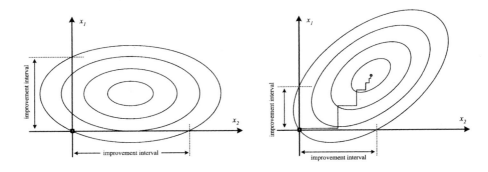

Fig. D.1. Left: a quadratic function that is aligned with the coordinate system has a relatively large improvement interval. Right: a rotation of the same function, which leads to much smaller improvement intervals. The difference increases as the eccentricity increases.

Thus, in order to estimate the maximal complexity of the DE algorithm, we should make it invariant to the axes' rotation. This gives us the real algorithm performance which does not depend on function characteristics (separable or epistatic ones).

Differentiation is rotationally invariant, because it is based on linear operators (2.8) that are invariant to the rotation of coordinate axes. In return, crossover is rotationally dependent. The outcome of (combinatorial) crossover is a vertex of a hypercube built on the trial and target individuals. Although the trial and target positions are invariant to the axes' rotation, the hypercube completely depends on it (see Fig. D.2). So, in order to make DE rotationally invariant we should exclude the crossover operation.

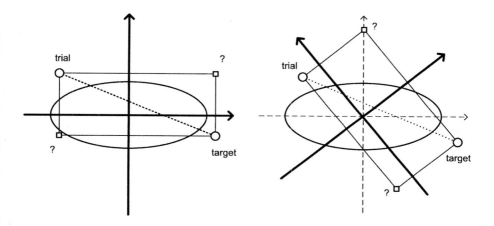

Fig. D.2. Left: coordinate and principal axes are aligned. Right: coordinate and principal axes are unaligned (rotated). Coordinate rotation shifts the location of potential child vectors generated by crossover.

Many experiments showed that a big value of crossover $Cr \approx 1 - 1/D$, that is, the trial individual inherits about one gene of the target individual at a time, ameliorates the convergence rate of separable functions significantly. In this case, such a crossover can be associated with a small rate mutation effect as mentioned above. In other words, crossover executes the function of D linear searchers independently for each variable. However, the real-life functions are usually epistatic ones, so such a speed-up feature of crossover is often idle.

E

Practical Guidelines to Application

In this appendix I briefly summarized all the practical information concerning the tuning of differential evolution for solving a problem. The tuning process is divided in five step-by-step phases.

1. *Type of strategy*
   If possible, you should use the knowledge of some characteristics of an objective function. For example, the group of directed strategies gives the best fit to "quasi" convex functions, and random strategies are preferable for difficult functions, where either we do not know or simply disregard their properties.
   Recommended values: Rand3 and Rand2/Best.

2. *Size of population*
   It should be not too small in order to avoid stagnation or local optima. On the other hand, the larger the size is, the more computations are needed.
   Recommended values: $NP = 10 \cdot D$.

3. *Constant of differentiation*
   $F \in (-1, 0) \cup (0, 1]+$ is proportional to the diversity of a population. Negative values localize the search between two barycenters of the difference vector. Positive values globalize the search.
   Recommended values: $F = 0.85$.

4. *Constant of crossover*
   $Cr \in [0, 1)$ adds a constructive diversity to the search process. It allows us to exploit some properties of a function such as the symmetry and separability.
   Recommended values: $Cr = 0.5$.

5. *Number of traversal steps*
   It is suggested to use the transversal DE in two cases: either to organize refined control of the diversity or for flexible parallel implementation of the algorithm.
   Recommended values: $n = 1$.

References

[Abb02]   Hussein A. Abbass. The self-adaptive Pareto differential evolution algorithm. In *Proceedings of the IEEE Congress on Evolutionary Computation, CEC2002*, pages 831–836, 2002.

[AD03]   Hussein A. Abbass and Kalyanmoy Deb. Searching under multi-evolutionary pressures. In *Proceedings of the 2003 Evolutionary Multiobjective Optimization Conference – EMO03*, pages 391–404. LNCS2632, Berlin, Springer-Verlag, 2003.

[Ang98]   Peter J. Angeline. Evolutionary optimization versus particle swarm optimization: Philosophy and performance differences. In V. W. Porto, N. Saravanan, D. Waagen, and A. E. Eiben, editors, *Evolutionary Programming VII*, pages 601–610, LNCS 1447, Berlin, Springer, 1998.

[AS02]   Hussein A. Abbass and Ruhul Sarker. A pareto differential evolution algorithm. *International Journal on Artificial Intelligence Tools*, 11(4):531–552, World Scientific, Singapore, 2002.

[ASN01a]   Hussein A. Abbass, Ruhul Sarker, and Charles Newton. A pareto differential evolution approach to vector optimisation problems. In *Proceedings of the IEEE Congress on Evolutionary Computation, CEC2001*, Seoul, Korea, IEEE Press, 2001.

[ASN01b]   Hussein A. Abbass, Ruhul Sarker, and Charles Newton. PDE: A pareto-frontier differential evolution approach for multi-objective optimization problems. In *Proceedings of the IEEE Congress on Evolutionary Computation, CEC2001*, pages 971–978, Seoul, Korea, IEEE Press, 2001.

[AT00]   M. Ali and A. Törn. Optimization of carbon and silicon cluster by differential evolution. In C.A. Floudas and P. Pardalos, editors, *Optimization in Computational Chemistry and Molecular Biology*, pages 287–300. Kluwer Academic, 2000.

[AT02]   M.M. Ali and A. Törn. Topographical differential evolution using pre-calculated differentials. In G. Dzemyda et al., editor, *Stochastic Methods in Global Optimization*, pages 1–17. Kluwer Academic, 2002.

[BD99]   H.G. Bayer and K. Deb. On the analysis of self adaptive evolutionary algorithms. Technical Report CI-69/99, University of Dortmund, 1999.

[Bey98]   Hans-Georg Beyer. On the explorative power of ES/EP-like algorithms. In V. W. Porto, N. Saravanan, D. Waagen, and A. E. Eiben, editors, *Evolutionary Programming VII*, pages 323–334, LNCS 1447, Berlin, Springer, 1998.

[BP95]   G. Bilchev and I. Parmee. The ant colony methaphor for searching continuous design space. In *Proceedings of AISB Workshop on Evolutionary Computation*, University of Sheffield, UK, 3–4 April 1995.

[BP96]   G. Bilchev and I. Parmee. Constrainted optimisation with an ant colony search model. In *Proceedings of ACED'96*, PEDC, University of Plymouth, UK, 1996.

[CL99]   F. Cheong and R. Lai. Designing a hierarchical fuzzy logic controller using differential evolution. In *Proceedings of 1999 IEEE International Fuzzy Systems Conference, FUZZ-IEEE'99*, volume 1, pages 277–282, Seoul, Korea, IEEE, Piscataway, NJ, 1999.

[Coe99a]   Carlos A. Coello Coello. A survey of constraint handling techniques used with evolutionary algorithms. Technical report, Laboratorio Nacional de Informática Avanzada, Xalapa, Veracruz, México, 1999. Lania-RI-99-04.

[Coe99b]   Carlos A. Coello Coello. The use of a multiobjective optimization technique to handle constraints. In Alberto A. Ochoa Rodríguez, Marta R. Soto Ortiz, and Roberto Santana Hermida, editors, *Proceedings of the Second International Symposium on Artificial Intelligence (Adaptive Systems)*, pages 251–256, Institute of Cybernetics, Mathematics and Physics, Ministry of Science, Technology and Environment, La Habana, Cuba, 1999.

[Coe02]   Carlos A. Coello Coello. Theoretical and numerical constraint-handling techniques used with evolutionary algorithms: A survey of the state of the art. In *Computer Methods in Applied Mechanics and Engineering*, volume 191, pages 1245–1287, January 2002.

[CW98]   Ji-Pyng Chiou and Feng-Sheng Wang. A hybrid method of differential evolution with application to optimal control problems of a bioprocess system. In *Proceedings of IEEE International Conference on Evolutionary Computation. IEEE World Congress on Computational Intelligence*, pages 627–632, New York, 1998.

[CW99]   Ji-Pyng Chiou and Feng-Sheng Wang. Hybrid method of evolutionary algorithms for static and dynamic optimization problems with application to a fed-batch fermentation process. *Computers and Chemical Engineering*, 23(9):1277–1291, November 1999.

[CWS01]   I.L. Lopez Cruz, L.G. Van Willigenburg, and G. Van Straten. Parameter control strategy in differential evolution algorithm for optimal control. In M.H. Hamza, editor, *Proceedings of the IASTED International Conference Artificial Intelligence and Soft Computing (ASC 2001)*, pages 211–216, Cancun, Mexico, ACTA, Calgary, 2001.

[CX00]   C.S. Chang and D.Y. Xu. Differential evolution based tuning of fuzzy automatic train operation for mass rapid transit system. In *IEE Proceedings on Electric Power Applications*, volume 147, pages 206–212, May 2000.

[CXQ99]   C.S. Chang, D.Y. Xu, and H.B. Quek. Pareto-optimal set based multiobjective tuning of fuzzy automatic train operation for mass transit system. In *IEE Proceedings on Electric Power Applications*, volume 146, pages 577–583, September 1999.

[Deb00]   Kalyanmoy Deb. An efficient constraint handling method for genetic algorithms. *Computer Methods in Applied Mechanics and Engineering*, 186(2–4):311–338, 2000.

[Deb01]   Kalyanmoy Deb. *Multi-Objective Optimization Using Evolutionary Algorithms*. Wiley, Chichester, UK, 2001.

[DeJ75]   Kenneth.A. DeJong. *The Analysis of the Behavior of a Class of Genetic Adaptive Systems*. PhD thesis, University of Michigan, Ann Harbor, University Microfilms No 76-9381, 1975.

[DJA01]   Kalyanmoy Deb, Dhiraj Joshi, and Ashish Anand. Real-coded evolutionary algorithms with parent-centric recombination. Technical Report 2001003, Kanpur Genetic Algorithms Laboratory (KanGAL), Department of Mechanical Engineering, Indian Institute of Technology, Kanpur, PIN 208 016, India, 2001.

[DPST03]   J. Dréo, A. Pétrowski, P. Siarry, and E. Taillard. *Métaheuristiques pour l'optimisation difficile*. Eyrolles, 2003.

[EBK98]   M.A. El-Beltagy and A.J. Keane. Optimisation for multilevel problems: A comparison of various algorithms. In I.C. Parmee, editor, *Adaptive Computing in Design and Manifacture*, pages 111–120. Springer-Verlag London Limited, 1998.

[EHM99]   Ágoston Endre Eiben, Robert Hinterding, and Zbigniew Michalewicz. Parameter control in evolutionary algorithms. *IEEE Trans. on Evolutionary Computation*, 3(2):124–141, 1999.

[ES03]   A.E. Eiben and J.E. Smith. *Introduction to Evolutionary Computing*. Springer, 2003.

[FF97]   C. M. Fonseca and Peter J. Fleming. Multiobjective optimization. In Thomas Bäck, David B. Fogel, and Zbigniew Michalewicz, editors, *Handbook of Evolutionary Computation*, pages C4.5:1–9. Institute of Physics Publishing and Oxford University Press, Bristol, New York, 1997.

[FJ03]   Vitaliy Feoktistov and Stefan Janaqi. Gestion de mission des satellites d'observation avec l'évolution différentielle. In *5-ème Congrès de la Société Française de Recherche Opérationnelle et d'Aide à la Décision - ROADEF 2003*, pages 228–230, Unitersité d'Avignon et des Pays de Vaucluse, 26–28 Février 2003.

[FJ04a]   Vitaliy Feoktistov and Stefan Janaqi. Classical identification problem solved by differential evolution: Choquet integral. In R. Matousek and P. Osmera, editors, *10th International Conference on Soft Computing – MENDEL 2004*, pages 62–67, Brno, Czech Republic, 16–18 June 2004.

[FJ04b]   Vitaliy Feoktistov and Stefan Janaqi. Differential evolution. Technical report, LGI2P - l'École des Mines d'Alès, Parc Scientifique G. Besse, 30035 Nîmes, France, January 2004.

[FJ04c]   Vitaliy Feoktistov and Stefan Janaqi. Evolution différentielle – une vue d'ensemble. In A. Dolgui and S. Dauzère-Pérès, editors, *Actes de la 5-ème Conférence Francophone de MOdélisation et SIMulation. Modélisation et simulation pour l'analyse et l'optimisation des systèmes industriels et logistiques – MOSIM 2004*, volume 1, pages 223–230, Nantes, France, 1–3 September 2004.

[FJ04d]   Vitaliy Feoktistov and Stefan Janaqi. Generalization of the strategies in differential evolution. In *18-th Annual IEEE International Parallel and Distributed Processing Symposium. IPDPS – NIDISC 2004 workshop*, page 165 and CD version, Santa Fe, NM, IEEE Computer Society, New York, 2004.

[FJ04e]   Vitaliy Feoktistov and Stefan Janaqi. Hybridization of differential evolution with least-square support vector machines. In *Proceedings of the Annual Machine Learning Conference of Belgium and The Netherlands – BENELEARN 2004*, pages 53–57, Vrije Universiteit Brussels, Belgium, 8–9 January 2004.

[FJ04f]   Vitaliy Feoktistov and Stefan Janaqi. New energetic selection principle in differential evolution. In *6th International Conference on Enterprise Information Systems – ICEIS 2004*, volume 2, pages 29–35, Universidade Portucalense, Porto – Portugal, 14–17 April 2004.

[FJ04g]   Vitaliy Feoktistov and Stefan Janaqi. New strategies in differential evolution. In Ian Parmee, editor, *6th International Conference on Adaptive Computing in Design and Manufacture – ACDM 2004*, pages 335–346, Bristol, UK, 20–22 April, Engineers House, Clifton, Springer-Verlag Ltd.(London) 2004.

[FJ04h]   Vitaliy Feoktistov and Stefan Janaqi. Transversal differential evolution : Comparative study. In Dana Petcu, Daniela Zaharie, Viorel Negru, and Tudor Jebelean, editors, *The 6th International Symposium on Symbolic and Numeric Algorithms for Scientific Computation – SYNASC 2004*, pages 490–501, Timisoara, Romania, 26–30 September 2004.

[FJ06]   Vitaliy Feoktistov and Stefan Janaqi. *Enterprise Information Systems VI*, chapter Artificial Intelligence and Decision Support Systems, pages 151–157. Springer, 2006.

[FL01]   Hui-Yuan Fan and Jouni Lampinen. A trigonometric mutation approach to differential evolution. In K. C. Giannakoglou, D. T. Tsahalis, J. Périaux, K. D. Papailiou, and T. Fogarty, editors, *Evolutionary Methods for Design Optimization and Control with Applications to Industrial Problems*, pages 65–70, Athens, Greece. International Center for Numerical Methods in Engineering (Cmine), 2001.

[FL03]   Hui-Yuan Fan and Jouni Lampinen. A directed mutation operation for the differential evolution algorithm. *International Journal of Industrial Engineering: Theory, Applications and Practice*, 1(10):6–15, 2003.

[Fle80]   Roger Fletcher. *Practical Methods of Optimisation, volume 1: Unconstrained Optimization*. John Wiley and Sons, Chichester, 1980.

[Fle81]   Roger Fletcher. *Practical Methods of Optimisation, volume 2: Constrained Optimization*. John Wiley and Sons, Chichester, 1981.

[Fog92]   D.B. Fogel. *Evolving Artificial Intelligence*. PhD thesis, University of California, San Diego, 1992.

[FOW66]   L.J. Fogel, A.J. Owens, and M.J. Walsh. *Artificial Intelligence Through Simulated Evolution*. John Wiley, New York, 1966.

[FPA$^+$99]   C.A. Floudas, P.M. Pardalos, C.S. Adjiman, W.R. Esposito, Z. Gumus, S.T. Harding, J.L. Klepeis, C.A. Meyer, and C.A. Schweiger. *Handbook of Test Problems for Local and Global Optimization*. Kluwer Academic, 1999.

[GHB98]   M.R. Ghasemi, E. Hinton, and S. Bulman. Performance of genetic algorithms for optimisation of frame stractures. In I.C. Parmee, editor, *Adaptive Computing in Design and Manufacture*, pages 287–299. Springer-Verlag London Limited, 1998.

[Glo90]   Fred Glover. Tabu search: A tutorial. *Interfaces*, 20(4):74–94, 1990.

[Gol89]   D.E. Goldberg. *Genetic Algorithms in Search, Optimization and Machine Learning*. Reading, MA, Addison Wesley, 1989.

[Gra97]   Michel Grabisch. k-ordered discrete fuzzy measures and their representation. *Fuzzy Sets and Systems*, (92):167–189, 1997.

[Hen]   Tim Hendtlass. A combined swarm differential evolution algorithm for optimization problems.

[Hol75]   J.H. Holland. *Adaptation in Natural and Artificial Systems*. University of Michigan Press, Ann Arbor, 1975.

[HW02]   Hsuan-Jui Huang and Feng-Sheng Wang. Fuzzy decision-making design of chemical plant using mixed-integer hybrid differential evolution. *Computers and Chemical Engineering*, 26(12):1649–1660, 2002.

[IKL03]   Jarmo Ilonen, Joni-Kristian Kamarainen, and Jouni Lampinen. Differential evolution training algorithm for feed-forward neural networks. *Neural Processing Letters*, 17(1):93–105, 2003.

[KE95]   J. Kennedy and R. C. Eberhart. Particle swarm optimization. In *Proc. of the IEEE Int. Conf. on Neural Networks*, pages 1942–1948, Piscataway, NJ, IEEE Service Center, 1995.

[Kea95]   Andy J. Keane. Genetic algorithm optimization of multi-peak problems: studies in convergence and rubustness. *Artificial Intelligence in Engineering*, 9(2):75–83, 1995.

[Kea96]   Andy J. Keane. A brief comparison of some evolutionary optimization methods. In V. Rayward-Smith, I. Osman, C. Reeves, G.D. Smith, and J. Wiley, editors, *Modern Heuristic Search Methods*, pages 255–272, 1996.

[KGV83]   S. Kirkpatrick, C. D. Gelatt Jr., and M. P. Vecchi. Optimization by simulated annealing. *Science*, 220(4598):671–680, 1983.

[KL04]   Saku Kukkonen and Jouni Lampinen. Mechanical component design for multiple objectives using generalized differential evolution. In Ian Parmee, editor, *6th International Conference on Adaptive Computing in Design and Manufacture – ACDM 2004*, pages 261–272, Bristol, UK, Engineers House, Clifton, Springer-Verlag Ltd.(London), 20–22 April 2004.

[KL05]   Saku Kukkonen and Jouni Lampinen. GDE3: The third evolution step of generalized differential evolution. In *The 2005 IEEE Congress on Evolutionary Computation*, 2005.

[Koz92]   John R. Koza. *Genetic Programming: On the Programming of Computers by Means of Natural Selection*. MIT Press, Cambridge, MA, 1992.

[KSL04]   Saku Kukkonen, Jouni Sampo, and Jouni Lampinen. Applying generalized differential evolution for scaling filter design. In R. Matousek and P. Osmera, editors, *10th International Conference on Soft Computing – MENDEL 2004*, pages 28–32, Brno, Czech Republic, 16–18 June 2004.

[Lam99]   Jouni Lampinen. Differential evolution - new naturally parallel approach for engineering design optimization. In Barry H.V. Topping, editor, *Developments in Computational Mechanics with High Performance Computing*, pages 217–228, Edinburgh, Civil-Comp Press, 1999.

[Lam01]   Jouni Lampinen. Solving problems subject to multiple nonlinear constraints by the differential evolution. In Radek Matousek and Pavel Osmera, editors, *Proceedings of MENDEL'01 – 7th International Conference on Soft Computing*, pages 50–57, Brno, Czech Republic, 6–8 June 2001.

[Lam02a]   Jouni Lampinen. A bibliography of differential evolution algorithm. Technical report, Department of Information Technology, Laboratory of Information Processing. Lappeenranta University of Technology, 2002.

[Lam02b]   Jouni Lampinen. A constraint handling approach for the differential evolution algorithm. In David B. Fogel, Mohamed A. El-Sharkawi, Xin Yao, Garry Greenwood, Hitoshi Iba, Paul Marrow, and Mark Shackleton, editors, *Proceedings of the 2002 Congress on Evolutionary Computation – CEC2002*, pages 1468–1473. Piscataway, NJ, IEEE Press, 2002.

[Las03]   Alexey Lastovetsky. *Parallel Computing on Heterogeneous Networks*. John Wiley and Sons, 2003. 423 pages.

[LHW00]   Yung-Chien Lin, Kao-Shing Hwang, and Feng-Sheng Wang. Plant scheduling and planning using mixed-integer hybrid differential evolution with multiplier updating. In *Proceedings of the CEC00, 2000 Congress on Evolutionary Computation*, volume 1, pages 593–600, Piscataway, NJ, IEEE, 2000.

[LL02a]   Junhong Liu and Jouni Lampinen. Adaptive parameter control of differential evolution. In P. Osmera, editor, *Proceedings of MENDEL'02 – 8th International Mendel Conference on Soft Computing*, pages 19–26, Brno, Czech Republic, 5–7 June 2002.

[LL02b]   Junhong Liu and Jouni Lampinen. On setting the control parameter of the differential evolution method. In P. Osmera, editor, *Proceedings of MENDEL'02 – 8th International Mendel Conference on Soft Computing*, pages 11–18, Brno, Czech Republic, 5–7 June 2002.

[LN89]   D.C. Lui and J. Nocedal. On the limited memory BFGS method for large scale optimization. *Mathematical Programming*, 45:503–528, 1989.

[LZ99a]   Jouni Lampinen and Ivan Zelinka. Mechanical engineering design optimization by differential evolution. In David Corne, Marco Dorigo, and Fred Glover, editors, *New Ideas in Optimization*, pages 127–146. London, McGraw-Hill, 1999.

[LZ99b]   Jouni Lampinen and Ivan Zelinka. Mixed integer-discrete-continuous optimization by differential evolution, Part 1: the optimization method. In P. Osmera, editor, *Proceedings of MENDEL'99 – 5th International Mendel Conference on Soft Computing*, pages 71–76, Brno, Czech Republic, 9–12 June 1999.

[LZ99c]   Jouni Lampinen and Ivan Zelinka. Mixed integer-discrete-continuous optimization by differential evolution, Part 2: A practical example.

In P. Osmera, editor, *Proceedings of MENDEL'99 – 5th International Mendel Conference on Soft Computing*, pages 77–81, Brno, Czech Republic, 9–12 June 1999.

[LZ00]   Jouni Lampinen and Ivan Zelinka. On stagnation of the differential evolution algorithm. In P. Osmera, editor, *Proceedings of MENDEL'00 – 6th International Mendel Conference on Soft Computing*, pages 76–83, Brno, Czech Republic, 7–9 June 2000.

[MA03]   N.P. Moloi and M.M. Ali. An iterative global optimization algorithm for potential energy minimization. Technical report, University of Minnesota, 22 March 2003.

[Mad02]   N.K. Madavan. Multiobjective optimization using a pareto differential evolution approach. In *The IEEE Congress on Evolutionary Computation*, pages 1145–1150, 2002.

[MBST99]   Javier G. Marin-Blàzquez, Qiang Shen, and Andrew Tuson. Tuning fuzzy membership functions with neighbourhood search techniques: A comparative study. In *Proceedings of the 3rd IEEE International Conference on Intelligent Engineering Systems*, pages 337–342, November 1999.

[MCTM04]   Efrén Mezura Montes, Carlos A. Coello Coello, and Edy I. Tun-Morales. Simple feasibility rules and differential evolution for constrained optimization. In Raúl Monroy, Gustavo Arroyo-Figueroa, Luis Enrique Sucar, and Humberto Sossa, editors, *Proceedings of the Third Mexican International Conference on Artificial Intelligence (MICAI'2004)*, volume 2972, pages 707–716. New York, Springer Verlag, LNAI, April 2004.

[MF02]   Zbigniew Michalewicz and David B. Fogel. *How to Solve It: Modern Heuristics*. Springer-Verlag, New York, 2002.

[Mic97]   Zbigniew Michalewicz. Other constraint-handling methods. In Thomas Bäck, David B. Fogel, and Zbigniew Michalewicz, editors, *Handbook of Evolutionary Computation*, pages C5.6:1–4. Institute of Physics Publishing and Oxford University Press, Bristol, 1997.

[MM05]   Rui Mendes and Arvind Mohais. DynDE: A Differential Evolution for dynamic optimization problems. In *The 2005 IEEE Congress on Evolutionary Computation*, 2005.

[MPV01]   G.D. Magoulas, V.P. Plagianakos, and M.N. Vrahatis. Hybrid methods using evolutionary algorithms for on-line training. In *Proceedings of IJCNN'01, International Joint Conference on Neural Networks*, volume 3, pages 2218–2223, Washington, DC, 15–19 July 2001.

[MS96]   Zbigniew Michalewicz and Marc Schoenauer. Evolutionary algorithms for constrained parameter optimization problems. *Evolutionary Computation*, 4(1):1–32, 1996.

[NM65]   J.A. Nelder and R. Mead. A simplex method for function minimization. *Computer Journal*, 7:308–313, 1965.

[NW88]   G.L. Nemhauser and L.A. Wolsey. *Integer and Combinatorial Optimization*. New York, John Wiley and Sons, 1988.

[NW99]   Jorge Nocedal and Stephen J. Wright. *Numerical Optimization*. Springer series in operations research. New York, Springer-Verlag, 1999.

[OES05] Mahamed G.H. Omran, Andries P. Engelbrecht, and Ayed Salman. Differential Evolution methods for unsupervised image classification. In *The 2005 IEEE Congress on Evolutionary Computation*, 2005.

[Pen04] Kalin Penev. Adaptive computing in support of traffic management. In Ian Parmee, editor, *6th International Conference on Adaptive Computing in Design and Manufacture – ACDM 2004*, pages 295–306, Bristol, UK, Engineers House, Clifton, Springer-Verlag Ltd.(London), 20–22 April 2004.

[PL03] K. Penev and G. Littlefair. Free Search – a novel heuristic method. In *Proceedings of the PREP 2003*, pages 133–134, Exeter, UK, 14–16 April 2003.

[PR02] P.M. Pardalos and E. Romeijn. *Handbook of Global Optimization – Volume 2: Heuristic Approaches*. Kluwer Academic Publishers, 2002.

[Pri94] Kenneth V. Price. Genetic annealing. *Dr. Dobb's Journal*, pages 127–132, October 1994.

[Pri97] Kenneth V. Price. Differential evolution vs. the functions of the 2nd ICEO. In *Proceedings of 1997 IEEE International Conference on Evolutionary Computation (ICEC '97)*, Cat. No.97TH8283, pages 153–157, Indianapolis, IN. IEEE; IEEE Neural Network Council (NNC); Evolutionary Computation (ICEC '97), 13-16 April 1997.

[Pri99] Kenneth V. Price. *New Ideas in Optimization*, chapter Differential Evolution. London, McGraw-Hill, 1999.

[PS82] C.H. Papadimitrou and K. Steiglitz. *Combinatorial Optimization: Algorithms and Complexity*. Englewood Cliffs, NJ, Prentice Hall, 1982.

[PS97] Kenneth Price and Rainer Storn. Differential evolution: A simple evolution strategy for fast optimization. *Dr. Dobb's Journal of Software Tools*, 22(4):18–24, April 1997.

[PSG+03] K. Pelckmans, J.A.K. Suykens, T.Van Gestel, J.De Brabanter, L. Lukas, B. Hamers, B.De Moor, and J. Vandewalle. LS-SVMlab Toolbox User's Guide. Technical Report 02-145, Katholieke Universiteit Leuven, Belgium, February 2003.

[PSL05] Kenneth V. Price, Rainer M. Storn, and Jouni A. Lampinen. *Differential Evolution: A Practical Approach to Global Optimization*. Natural Computing Series. New York, Springer, 2005.

[PTP+04] K.E. Parsopoulos, D.K. Tasoulis, N.G. Pavlidis, V.P. Plagianakos, and M.N. Vrahatis. Vector evaluated differential evolution for multiobjective optimization. In *Proceedings of the 2004 IEEE Congress on Evolutionary Computation*, pages 204–211, Portland, OR, Piscataway, NJ, IEEE Press, 2004.

[QS05] A.K. Qin and P.N. Suganthan. Self-adaptive Differential Evolution algorithm for numerical optimization. In *The 2005 IEEE Congress on Evolutionary Computation*, 2005.

[RD00] T. Rogalsky and R.W. Derksen. Hybridization of differential evolution for aerodynamic design. In *Proceedings of the 8th Annual Conference of the Computational Fluid Dynamics Society of Canada*, pages 729–736, June 11–13 2000.

[RDK99a] T. Rogalsky, R.W. Derksen, and S. Kocabiyik. An aerodynamic design technique for optimizing fan blade spacing. In *Proceedings of the 7th Annual Conference of the Computational Fluid Dynamics Society of Canada*, pages 2–29 – 2 –34, 30 May – 1 June 1999.

[RDK99b]   T. Rogalsky, R.W. Derksen, and S. Kocabiyik. Differential evolution
in aerodynamic optimization. In *Proceedings of the 46th Annual Conference of the Canadian Aeronautics and Space Institute*, pages 29–36,
2–5 May 1999.

[RDK99c]   T. Rogalsky, R.W. Derksen, and S. Kocabiyik. Optimal optimization
in aerodynamic design. In *Proceedings of the 17th Canadian Congress
of Applied Mechanics*, 30 May – 3 June 1999.

[Rec73]   I. Rechenberg. *Evolutionstrategie: Optimierung Technisher Systeme
nach Prinzipien des Biologischen Evolution*. Fromman-Hozlboog Verlag, Stuttgart, 1973.

[RKD00]   T. Rogalsky, S. Kocabiyik, and R.W. Derksen. Differential evolution in aerodynamic optimization. *Canadian Aeronautics and Space
Journal*, 46(4):183–190, December 2000.

[RL03]   Jani Rönkkönen and Jouni Lampinen. On using normally distributed mutation step length for the differential evolution algorithm. In
R. Matoušek and P. Ošmera, editors, *Proc. of MENDEL'03 – 9th International Conference on Soft Computing*, pages 11–18, Brno, Czech
Republic, 4–6 June 2003.

[RM01]   James A. Rumpler and Frank W. Moore. Automatic selection of subpopulation and minimal spanning distances for improved numerical
optimization. In *Proceedings of the IEEE Congress on Evolutionary
Computation, CEC2001*, 2001.

[Rog98]   T. Rogalsky. Aerodynamic shape optimization of fan blades. Master's
thesis, University of Manitoba. Department of Applied Mathematics,
1998.

[RS05]   Krzysztof Rzadca and Franciszek Seredynski. Heterogeneous multiprocessor scheduling with Differential Evolution. In *The 2005 IEEE
Congress on Evolutionary Computation*, 2005.

[Rüt97a]   Martin Rüttgers. Design of a new algorithm for scheduling in parallel
machine shops. In *Proceedings of the 5th European Congress on Intelligent Techniques and Soft Computing*, volume 3, pages 2182–2187,
1997.

[Rüt97b]   Martin Rüttgers. Differential evolution: A method for optimization of
real scheduling problems. Technical report, TR-97-013, International
Computer Science Institute, 1997.

[SA04]   Ruhul Sarker and Hussein A. Abbass. Differential evolution for solving multiobjective optimization problems. *Asia-Pacific Journal of
Operational Research*, 21(2):225–240, 2004.

[Saa80]   Thomas L. Saaty. *The Analytic Hierarchy Process*. New York,
McGraw-Hill, 1980.

[Sal96]   Ralf Salomon. Re-evaluating genetic algorithm performance under
coordinate rotation of benchmark functions: A survey of some theoretical and practical aspects of genetic algorithms. *BioSystems*,
39(3):263–278, 1996.

[Sal00]   Michel Salomon. Parallélisation de l'évolution différentielle pour le re-
calage rigide d'images médicales volumiques. In *RenPar'2000, 12ème
Rencontres Francophones du Parallélisme*, Besançon (France), 19–22
Juin 2000.

[Sch81]   H.P. Schwefel. *Numerical Optimization of Computer Models*. New
York, John Wiley and Sons, 1981. 1995, second edition.

[SDB$^+$93]　William M. Spears, Kenneth A. De Jong, Thomas Bäck, David B. Fogel, and Hugo de Garis. An overview of evolutionary computation. In Pavel B. Brazdil, editor, *Proc. of the European Conf. on Machine Learning*, pages 442–459, Berlin, Springer, 1993.

[SGB$^+$02]　J.A.K. Suykens, T.Van Gestel, J.De Brabanter, B.De Moor, and J. Vandewalle. *Least Squares Support Vector Machines*. Singapore, World Scientific, 2002.

[SHH62]　W. Spendley, G.R. Hext, and F.R. Himsworth. Sequential application of simplex designs in optimisation and evolutionary operation. *Technometrics*, 4:441–461, 1962.

[SM96]　Marc Schoenauer and Zbigniew Michalewicz. Evolutionary computation at the edge of feasibility. In Hans-Michael Voigt, Werner Ebeling, Ingo Rechenberg, and Hans-Paul Schwefel, editors, *Parallel Problem Solving from Nature – PPSN IV*, pages 245–254, Berlin, Springer, 1996.

[SMdMOC03]　Hélder Santos, José Mendes, P.B. de Moura Oliveira, and J. Boaventura Cunha. Path planning optimization using the differential evolution algorithm. In *Actas do Encontro Cientifico – 3° Festival Nacional de Robotica – ROBOTICA2003*, Lisboa, 9 de Maio de 2003.

[SP95]　Rainer Storn and Kenneth Price. Differential evolution – A simple and efficient adaptive scheme for global optimization over continuous spaces. Technical Report TR-95-012, International Computer Science Institute, Berkeley, CA, 1995.

[SP96]　Rainer Storn and Kenneth Price. Minimizing the real functions of the ICEC'96 contest by differential evolution. In *IEEE International Conference on Evolutionary Computation*, pages 842–844, Nagoya, IEEE, New York, May 1996.

[SP97]　Rainer Storn and Kenneth Price. Differential evolution – A simple and efficient heuristic for global optimization over continuous spaces. *Journal of Global Optimization*, (11):341–359, December 1997.

[Spe93]　William M. Spears. Crossover or mutation? In L. Darrell Whitley, editor, *Foundations of Genetic Algorithms 2*, pages 221–237. San Mateo, CA, Morgan Kaufmann, 1993.

[Spe98]　William M. Spears. *The Role of Mutation and Recombination in Evolutionary Algorithms*. PhD thesis, George Mason University, Fairfax, VA, 1998.

[SPH00]　M. Salomon, G.-R. Perrin, and F. Heitz. Parallelizing differential evolution for 3d medical image registration. Technical report, ICPS, Univ.-Strasbourg, September 2000.

[Sto95]　Rainer Storn. Differential evolution design of an IIR-filter with requirements for magnitude and group delay. Technical Report TR-95-026, International Computer Science Institute, Berkeley, CA, June 1995.

[Sto96a]　Rainer Storn. On the usage of differential evolution for function optimization. In *Biennial Conference of the North American Fuzzy Information Processing Society (NAFIPS 1996)*, pages 519–523, Berkeley, CA, New York, IEEE, 1996.

[Sto96b]　Rainer M. Storn. System design by constraint adaptation and differential evolution. Technical Report TR-96-039, International Computer Science Institute, Berkeley, CA, November 1996.

[Sto99]   Rainer Storn. Designing digital filters with differential evolution. In David Corne, Marco Dorigo, and Fred Glover, editors, *New Ideas in Optimization*, pages 109–125. London, McGraw-Hill, 1999.

[SV99]   J.A.K. Suykens and J. Vandewalle. Least squares support vector machines classifiers. *Neural Processing Letters*, 9(3):293–300, 1999.

[Swa72]   W. Swann. Direct search methods. In W. Murray, editor, *Numerical Methods for Unconstrained Optimization*, pages 13–28, New York, Academic, 1972.

[Tho03]   René Thomsen. Flexible ligand docking using differential evolution. In *Proceedings of the 2003 Congress on Evolutionary Computation*, volume 4, pages 2354–2361, 2003.

[Tho04]   René Thomsen. Multimodal optimization using crowding-based differential evolution. In *Proceedings of the 2004 Congress on Evolutionary Computation*, volume 2, pages 1382–1389, 2004.

[Tor89]   V. Torczon. *Multi-directional Search: A Direct Search Algorithm for Parallel Machines*. PhD thesis, Rice University, Huston, TX, 1989.

[TV97]   P. Thomas and D. Vernon. Image registration by differential evolution. In *Proceedings of the First Irish Machine Vision and Image Processing Conference IMVIP-97*, pages 221–225, Magee College, University of Ulster, 1997.

[Tvi04]   Josef Tvirdik. Generalized controlled random search and competing heuristics. In R. Matousek and P. Osmera, editors, *Mendel'04 – 10th International Conference on Soft Computing*, pages 228–233, Brno, Czech Republic, 16–18 June 2004.

[UV03]   Rasmus K. Ursem and Pierré Vadstrup. Parameter identification of induction motors using differential evolution. In *Proceedings of the 5th Congress on Evolutionary Computation, CEC2003*, volume 2, pages 790–796, 8–12 December 2003.

[vA92]   P. J. M. van Laarhoven and E. H. L. Aarts. *Simulated annealing: Theory and applications*. Dordrecht, Kluwer, 1992.

[Vap95]   Vladimir Vapnik. *The Nature of Statistical Learning Theory*. New-York, Springer-Verlag, 1995.

[VRSG00]   P. Vancorenland, C. De Ranter, M. Steyaert, and G. Gielen. Optimal rf design using smart evolutionary algorithms. In *Proceedings of 37th Design Automation Conference*, Los Angeles, 5–9 June 2000.

[Š02]   Tomislav Šmuc. Improving convergence properties of the differential evolution algorithm. In P. Osmera, editor, *Proceedings of MENDEL'02 – 8th International Mendel Conference on Soft Computing*, pages 80–86, Brno, Czech Republic, 5–7 June 2002.

[VT04]   J. Vesterstrøm and R. Thomsen. A comparative study of differential evolution, particle swarm optimization, and evolutionary algorithms on numerical benchmark problems. In *Proceedings of the 2004 Congress on Evolutionary Computation*, volume 2, pages 1980–1987, 2004.

[WC97a]   Feng-Sheng Wang and Ji-Pyng Chiou. Differential evolution for dynamic optimization of differential-algebraic systems. In *Proceedings of the IEEE International Conference on Evolutionary Computation – ICEC'97*, pages 531–536, Indianapolis, IN. Piscataway, NJ, IEEE Press, 1997.

184   References

[WC97b]   Feng-Sheng Wang and Ji-Pyng Chiou. Optimal control and optimal
             time location problems of differential-algebraic systems by differential
             evolution. *Ind. Eng. Chem. Res*, 36:5348–5357, 1997.

[wCzCzC02]   Chong wei Chen, De zhao Chen, and Guang zhi Cao. An improved dif-
             ferential evolution algorithm in training and encoding prior knowledge
             into feedforward networks with application in chemistry. *Chemomet-
             rics and Intelligent laboratory systems*, (64):27–43, 2002.

[WM95]   David H. Wolpert and William G. Macready. No free lunch theorems
             for search. Technical Report SFI-TR-95-02-010, Santa Fe Institute,
             Santa Fe, NM, July 1995.

[WM97]   David H. Wolpert and William G. Macready. No free lunch theo-
             rems for optimization. *IEEE Trans. on Evolutionary Computation*,
             1(1):67–82, 1997.

[Wri95]   Margaret H. Wright. Direct search methods: once scorned, now re-
             spectable. In D.F. Griffiths and G.A. Watson, editors, *Proceedings of
             the 1995 Dundee Biennial Conference in Numerical Analysis*, pages
             191–208, Harlow, Addison Wesley Longman, 1995.

[Wri96]   M.H. Wright. The Nelder–Mead method: Numerical experimentation
             and algorithmic improvements. Technical report, AT&T Bell Labo-
             ratories, Murray Hill, NJ, 1996.

[XSG03a]   Feng Xue, Arthur C. Sanderson, and Robert J. Graves. Multi-
             objective differential evolution and its application to enterprise plan-
             ning. In *The IEEE International Conference on Robotics and Au-
             tomation*, 2003.

[XSG03b]   Feng Xue, Arthur C. Sanderson, and Robert J. Graves. Pareto-based
             multi-objective differential evolution. In *The IEEE Congress on Evo-
             lutionary Computation*, 2003.

[XSG05a]   Feng Xue, Arthur C. Sanderson, and Robert J. Graves. Modeling
             and convergence analysis of a continuous multi-objective differential
             evolution algorithm. In *The 2005 IEEE Congress on Evolutionary
             Computation*, 2005.

[XSG05b]   Feng Xue, Arthur C. Sanderson, and Robert J. Graves. Multi-
             objective differential evolution – Algorithm, convergence, analysis and
             applications. In *The 2005 IEEE Congress on Evolutionary Computa-
             tion*, 2005.

[Xue04]   Feng Xue. *Multi-Objective Differential Evolution: Theory and Appli-
             cations*. PhD thesis, Rensselaer Polytechnic Institute, 2004.

[XZ04]   Xiao-Feng Xie and Wen-Jun Zhang. SWAF: Swarm algorithm frame-
             work for numerical optimization. In K. et al. Deb, editor, *Genetic
             and Evolutionary Computation Conference (GECCO) – Proceedings,
             Part I. LNCS 3102*, pages 238–250, Seattle, WA, New York, Springer-
             Verlag, 2004.

[Zah01]   Daniela Zaharie. On the explorative power of differential evolution
             algorithms. In *Proc. of SYNASC'2001 – Analele Univ. Timisoara*,
             volume XXXIX, pages 249–260, Timisoara, Roumania, 2001.

[Zah02]   Daniela Zaharie. Parameter adaptation in differential evolution by
             controlling the population diversity. In *Proc. of SYNASC'2002 –
             Analele Univ. Timisoara, seria Matematica-Informatica*, volume XL,
             special issue, pages 281–295, Timisoara, Roumania, 2002.

[Zah03]   Daniela Zaharie. Multi-objective optimization with adaptive pareto differential evolution. In *Memoriile Sectiilor Stiintifice, Seria IV, Tomul XXVI*, pages 223–239, 2003.

[Zah04]   Daniela Zaharie. A multi-population differential evolution algorithm for multi-modal optimization. In R. Matousek and P. Osmera, editors, *Mendel'04 – 10th International Conference on Soft Computing*, pages 17–22, Brno, Czech Republic, 16–18 June 2004.

[ZP03]   Daniela Zaharie and Dana Petcu. Adaptive pareto differential evolution and its parallelization. In *Fifth International Conference on Parallel Processing and Applied Mathematics*, pages 261–268, Czestochowa, Poland, September, LNCS 3019, 2003.

[ZX03]   Wen-Jun Zhang and Xiao-Feng Xie. DEPSO: Hybrid particle swarm with differential evolution operator. In *IEEE Int. Conf. on Systems, Man and Cybernetics*, pages 3816–3821, Washington DC, 2003.

Index

Abbass, 6
Ackley's function, 126, **163**
ACO, *see* ant colony optimization
action of algorithm, 112
adaptation
  schemes, 5
  size of population, 113
adaptive control, 76
aggregation formula, 133
aggregation problem, 134
AHP, *see* analytical hierarchy process
analogy, 5, 101–108
  free search, 108
  nonlinear simplex, 103
  particle swarm optimization, 105
analytical hierarchy process, 134
animal's behavior
  action, 107
  sence, 107
ant colony optimization, 2, 106
antecedent strategies, 41
applications, 12, 133–142
  decision-making, 133
  engineering design, 139
approximate methods, 8
approximating function, 123
approximation, 6, 78, **123**
  barycenter, 129
  optimum of, 124
    linear system, 125
  quality, 123
architecture of algorithm, 5
  sequential, 93

transversal, 94
two-array, 92
artificial evolution, 2
augmented system, 122
average population diversity, 5, **74**

badly conditioned matrix, **125**, 126
banana function, 10
barycenter, 44, **72**
base point, *see* base vector, **42**
base vector, 29, 42, **71**
  local, 42
  random, 42
benchmarks, 63, 95, *see* test function
Beyer's postulate, 5, **69**
BFGS algorithm, 11
binary-continuous DE, 5
bioinformatics, 13
biosystems, 13
boundary constraints, 28, **33**
  other techniques, 33
  periodic mode, 5
  shifting mechanism, 5
bump problem, 139
  best-known solutions, 139
  implementation of DE, 141

C language, 14, **149**, 157
chemical engineering, 13
child, 30
children, 26
Choquet Integral, 133, **135**
classical identification problem, 135
classical methods, 7

classification, 121
CMA-ES, 18
Coello Coello, 5
collective intelligence, 1, **106**
combinatorial optimization, 7
complexity of extra function, 125
computation chemistry, 13
conjugate gradient method, 122
conscious thinking imitation, 106
constraints, 5, **33**
    boundary, 28, **33**
        other techniques, 33
        periodic mode, 5
        reinitialization, 5
    inequality, 28
    modification of selection rules, 5, **34**
    other techniques, 35
    penalty method, 5, **34**
    Web site, 5
construction function, 70
construction theory, 70
continuous optimization, 7, **41**
    fundamental principle, 42
contraction, 102
contribution (personal), 7, **148**
control, 13
control parameters, 5, 14, 29, 73–77
    goal, 73
    influence, 75–76
    recommendations, 5, **171**
    robustness, 5, **86**
    self-adaptation, 77
    tuning, 76–77
        complexity of, 85
convergence
    improvement, 5, 7, 77–78, 96, 111
        energetic selection, 117, **118**
        hybridization, 128
    integrating criteria, 84
    measure, 84
    objective function, 86
    population, 87
    probability of, 84
    speed-up, 6
convex function, 78
coordinate rotation, 167
cost function, 28, *see* criterion of
        optimization
crafting effort, 85

criterion of optimization, 1, *see*
        objective function
critical generation, 126
crossover, 4, 26, **29**, 70–71
    applying to DE, 70
    arithmetic, 4
    binary, 4
    exponential, 4
    favourable effect, 18
    other approaches, 29
    principal role, 70
    rotationally dependent, 169
    small rate mutation effect, 169
crossover constant, **29**
    influence, 75
    self-adaptation, 5
crowding-based DE, 6

damping mechanisms, 105
DE, *see* differential evolution
Deb, 7
decision maker, 134
decision-making, 13
decreasing of population size, 5
DeJong, 25, **159**
deterministic control, 76
development, 4–7
difference vector, 4, 29, 42, **71**
    barycenter, 72
    decreasing, 71
    directed, 42
    random, 42
differences, 2
differential evolution, 145
    assumption, 42
    concept, 18
    development, 4–7
    famous algorithm, 13–16
    fundamental idea, 18
    history, 2–4
    keys of success, 20
    newest statement, 28
    performance (estimation), 168
    secrets of success, 17–20
    serves for, 7–13
differential mutation, 3, 71
    alternation with
        triangle scheme, 4, **42**
        weighted directed scheme, 4, **42**

directions, 4
  semi-directed, 4
  similarity, 71
differentiation, 4, 29, 41, **42**, 47, 70,
      106, 145
  analysis, 71
  barycenter, 42
  general operator, 70
  rotationally invariant, 169
  self-adaptation, 71
  type, 43
differentiation constant, **29**
  adaptation, 5
  fuzzy control, 5
  influence, 75
  negative, 46
  positive, 46
  range, 46
  relaxation, 5, **75**
  self-adaptation, 5
difficult optimization, 9
direct search methods, 101
directed scheme, 42
direction, 18, **42**
  factor, 73
  good, 43
discrete variables, 32
disruption theory, 70
diversity, 5, 70, **72**
  control, 147
    decreasing of population size, 5
    differentiation constant, 5, **70**
    precalculated differentials, 5
    refresh of population, 5
    strategy, 6
    transversality, 96
  decreasing, 96
  direct measures, 74
  estimation, 73
    average population, 5, **74**
    combinatorial approach, 4, **72**
    expected population variance, 5, **74**
    mean square, 5, **74**
    P-measure, 5, **75**
  extra, 96
  practical remark, 73
  transfer function, 74
double selection principle, 117

EA, *see* evolutionary algorithms
Eberhart, 104
EC, *see* evolutionary computation
Eiben, 26
elementary evolution step, 113
elitist selection, 19, **29**
energetic approach, 111–113
  action, 112
  potential energy, 112
energetic barriers, 113
  linear, 114
  nonlinear, 115
energetic filter, 6, **113**
energetic selection, 111–118
  advantage of, 116
  numerical results, 117–118
  practical remarks, 117
  principle, 113–116
    idea, 113
Enes-measure, 84
engineering design, 4, 12
engineering problems, 7
entropy, 85
ES, *see* evolution strategy
evolution, **1**, 25
evolution strategy, **26**, 71, 101
evolutionary algorithms, 2, **26**, 25–27,
      145
  basic scheme, 26
  general trends
    evolution strategies, 26
    evolutionary programming, 26
    genetic algorithms, 25
    genetic programming, 26
    typical EA, 27
evolutionary computation, **25**, 145
evolutionary cycle, 26
  of individual, 30
  three steps, 26
evolutionary programming, 26
exact methods, 8
examples of strategies, *see* strategies,
      47–63
  notation, 47
excessive
  exploitation, 69
  exploration, 69
    example, 95
expansion, 102

expected population variance, 5
exploitation, 77
exploitation function of strategy, 70
exploration, 43, 104
    careful, 116
exploration function of strategy, 70
exploration/exploitation, 69
    combinatoiral estimation, 73
extra function, 6, **123**
    four steps (summarized), 125

famous algorithm, 13–16
    four steps, 15
    graphical interpretation, 17
    source codes, 149–156
    summirized, 16
Fan, 4, 41
feasible points, 28
feature space, 122
Feoktistov, 148
fitness, 14, **28**
    evaluation, 26
Fogel D., 26
Fogel L., 26
forced stopping, 117
four groups of strategies, 4, 42–46
foxholes, *see* Shekel's function
free search, 5, 106–108, 139
    drawbacks, 108
    random factors, 108
FS, *see* free search
fuzzy systems, 13

GA, *see* genetic algorithms
general operator, 70
generate-and-test rules, 6
generation, **26**, 29, 113
genes, **26**, 28
genetic algorithms, 2, **25**, 71, 101, 139,
    168
genetic annealing, 3
genetic programming, 26
global methods, 7
global minimum, 28
global optimization, 2
Goldberg, 26
gradient simulation, 42
greedy selection, *see* elitist selection
group

RAND, 44
RAND/BEST, 45
RAND/BEST/DIR, 46
RAND/DIR, 44

Hendtlass, 6
heterogeneous networks of computers,
    5, 97–99
heuristic methods, 8
history, 2–4
Holland, 25
human cognition simulation, 104
hybridization, 121–129
    algorithm of, 125
    barycenter approximation, 129
    critical generation, 126
    efficiency measure, 129
    extra function, 125
    four levels
        external level, 6
        individual level, 6
        meta level, 6
        population level, 6
    idea, 123
    L-BFGS, 78
    numerical results, 125–128
    observations, 129
    two phases of, 125

idea of energetic selection, 113
identification (classical problem), 135
image processing, 13
imitation of a gradient, 43, *see*
    RAND/DIR
individual, 1, **26**, 28
    average normalized distance between,
        74
    best, 43
    cycle in DE, 30
    deterministic creation, 123
    injecting, 76
    number, 72, **76**
    potential, 111
    replacement, 78
    trial, **29**, 71
inertia weight, 105
influence constant, 45
initialization, 14, 26, **28**, 32
    large size, 116

integer variables, 32
intelligence of algorithm, 77
intelligent selection rules, 157–158
interpretation via DE, 101
    free search, 108
    nonlinear simplex, 103
    particle swarm optimization, 105
        complementarity, 106
iteration, 15, **26**

Karush–Kuhn–Tucker conditions, 122
Keane, 139
Kennedy, 104
kernel function, 78, 121, **123**
    examples, 124
Koza, 26
Kukkonen, 7

L-BFGS, 6, **78**, *see* BFGS algorithm
Lagrangian, 122
Lampinen, 4, 7, 32, **34**, 41
    Web site, 4
large-scale optimization, 6
Lastovetsky, 97
learning database (set), 136
least squares SVM, 122
levels of performances improvement
    external, 147
    individual, 146
    population, 147
limits, 87
linear energetic barriers, 114
linear programming, 7
linear system, 122
Littlefair, 106
local methods, **7**, 78
local minimum, 28
localization of promising zones, 78, **117**
LS-SVM, *see* least squares SVM
LS-SVMlab Toolbox, 126
Lui, 6

Madavan, 6
MATLAB, 10, 14, 126, 134, 142, **153**
Mead, 102
mean square diversity, 5
measure, 83–87, 146
    Enes, 84
    P-measure, 86

Q-measure, 83
    R-measure, 86
mechanical engineering, 33
memory aspect, 105
Mercer's condition, 121, **122**
metaheuristic methods
    advantage of, 9
metaheuristics
    classification
        neighborhood, 10
        population-based, 10
method, 145
    classical, 7
    direct search, 101
    global, 7
    local, 7
    mixed variables, 4
    regression, **123**, 148
migration concept, 6
minimal spanning distance, 6
minimization, 28
    of potential energy, 113
minimum location, 28
mixed variables, 4, 8, **32**
    boundary constraints, 33
    discrete, 32
    integer, 32
mixed-integer programming, 8
modification of selection rules, **34**, 157
molecular biology, 13
Moore, 6
mpC, *see* multiparallel C
multimodal optimization, 6
multiobjective optimization, 6
    single objective function via, 7
multiparallel C, 97
multipeak problems, 139
multipopulation optimization, 6
multistarts, 78
mutation, 26, **70**
    small probability of, 168
mutation parameter, 31

Nelder, 102
Nelder–Mead algorithm, 102
    advantage of DE, 103
    drawbacks, 103
neoteric differential evolution, 25–36
    advantage of, 31

algorithm, 29
  distinction of, 30
neural networks, 13
Newton, 6
No Free Lunch theorem, 43
Nocedal, 6
nonlinear energetic barriers, 115
nonlinear function estimation, 121
nonlinear mapping, 122
nonlinear programming, 7
nonlinear simplex, 102–104
  advantage of DE, 103
  drawbacks, 103

objective function, **28**, 43, 123
  convergence of, 86
operating variables, 14
operation, *see* operator
' dynamic effect, 71
operator
  capability of, 69
  crossover, 4, 18, 26, **29**, 70–71
  differentiation, 4, **29**, 70
  general, 70
  linear, 2
  mutation, 26, **70**
  relative importance, 69
  selection, 4, 19, **29**
optimality conditions, **122**, 123, *see*
  Karush–Kuhn–Tucker conditions
optimization, 28
  combinatorial, 7
    approximate methods, 8
    exact methods, 8
  continuous, 7
    linear programming, 7
    nonlinear programming, 7
    quadratic programming, 7
  difficult, 9
  general scheme, 7
  metaheuristic, 9
  mixed variables, 8
  multimodal, 6
  multiobjective, 6
    C-MODE and D-MODE, 7
    single objective function via, 7
  multipopulation, 6
  pareto, **7**, 77
  population-based, 10

problem definition, 28
traditional, 8
optimizer, 3
optimum
  approximation, 6, 78
  global, 28
  local, 28
  localization, 78

P-measure, 5, 75, *see* population
  measure
parallel computation, 97
parallelization, **97**, 147
  flexibility of transversal DE, 98
parameters control
  adaptive, 76
  deterministic, 76
  self-adaptive, 77
parents, 26
pareto-approach, 7
  selection rules, 34
  source code, 157
particle swarm optimization, 2, 5,
  104–106
  alternation with DE, 6
  combination with DE, 6
  premature convergence, 105
penalty function, 34
Penev, 106, **139**
performance loss, 168
performance measures, *see* measure
periodic mode, 5
pheromone, 107
population, 1, **26**, 28
  control, 76
  convergence, 87
  potential energy, 112
    minimization of, 113
  radius, 75
  reduction, 78
    considerable, 117
    dynamics of, 115
  refresh, 76
  size (influence), 75
population measure, 86
population-based optimization
  advantage of, 10
  examples, 10–11
  features, 11

potential difference, 112
potential energy, 112
  minimization, 113
potential of individual, 111
potential optimal solution, 28
practical guidelines, 171
precalculated differentials, 5, **97**
precision, 86
  improvement, 7
preselection, 4
Price, **3**, 13, 41, 85
probability of convergence, 84
problem definition, 28
problem parameters, 26
promising zones, 78, **117**
PSO, *see* particle swarm optimization

Q-measure, *see* quality measure
quadratic programming, **7**, 122
quality measure, 83
  advantage of, 84
Quartic function, 162

R-measure, *see* robustness measure
radius of population, *see* P-measure
Rand0/Best/Dir1, 59
Rand1, 47
Rand1/Best, 56
Rand1/Best/Dir1, 60
Rand1/Best/Dir2, 60
Rand1/Dir1, 51
Rand2, 48
Rand2/Best, 57
Rand2/Best/Dir1, 61
Rand2/Best/Dir3, 61
Rand2/Dir1, 51
Rand3, 48
Rand3/Best, 58
Rand3/Best/Dir4, 61
Rand3/Dir2, 53
Rand3/Dir3, 53
Rand4, 50
Rand4/Best, 58
Rand4/Best/Dir4, 63
Rand4/Best/Dir5, 63
Rand4/Dir2, 54
Rand4/Dir3, 54
Rand5, 50
Rand5/Dir4, 55

random number generator, **13**, 149
random subspaces, 103
Rastrigin's function, 11, 126, **163**
Rechenberg, 26
recombination, *see* crossover
  continuous, 71
reflection, 102
refresh of population, 5
regression methods, 78, **123**, 148
reinitialization, 5, **33**
relaxation of differentiation, 75
replacement, **26**, 126
reproduction, 29
robustness measure, 86
Rosenbrock's function, 10, 14, 117, **159**
rotated ellipsoid function, 117, 126, **164**
Rumpler, 6

Salomon, 168
Sarker, 6
scheduling, 13
scheme, *see* strategy
  DE/best/1, 41
  DE/best/2, 41
  DE/rand-to-best/1, 41
  DE/rand/1, 41
  DE/rand/2, 41
  directed, 42
  trigonometric, 42
Schwefel, 26
search
  behavior, 4, **43**
    animal, 106
    social, 106
  direct methods, 101
  global, 7
  inertia, 106
  local, 7
  random subspaces, 103
  step, 18
secrets of success, 17–20
selection, 4, 19, 26, **29**
  double, 117
  intelligent rules, 157
self-adaptation, 71
  control parameters, 77
self-adaptive control, 77
semi-directed mutation, 4
separable function, 167

sequential DE, 93
sharing scheme, 6
Shekel's function, 162
shift, 45
shifting mechanism, 5
shrinkage, 102
simple nostalgia rule, 104
single objective function, 7
size of population, 75
    control of, 76
Smith, 26
social influence rule, 104
social strategy, 106
source code, 14
    C language, 149
    MATLAB, 153
species, 5, **91**
    comparison of, 95–97
    sequential, 93
    transversal, **94**, 146
    two-array, 92
Sphere function, 117, **159**
SQP method, 138
stability of algorithm, 43
stagnation effect, 6, **35**
    possible, 117
standard test suite, 159–164
state of the art of DE, 7
steepest descent, 10
step function, 160
step length, 18
stopping condition, 15, **26**, 29
Storn, **3**, 13, 41
    Web site, 3
strategies, *see* strategy
    notation, 47
    RAND strategies, 47
        Rand1, 47
        Rand2, 48
        Rand3, 48
        Rand4, 50
        Rand5, 50
    RAND/BEST strategies, 56
        Rand1/Best, 56
        Rand2/Best, 57
        Rand3/Best, 58
        Rand4/Best, 58
    RAND/BEST/DIR strategies, 58
        Rand0/Best/Dir1, 59

        Rand1/Best/Dir1, 60
        Rand1/Best/Dir2, 60
        Rand2/Best/Dir1, 61
        Rand2/Best/Dir3, 61
        Rand3/Best/Dir4, 61
        Rand4/Best/Dir4, 63
        Rand4/Best/Dir5, 63
    RAND/DIR strategies, 51
        Rand1/Dir1, 51
        Rand2/Dir1, 51
        Rand3/Dir2, 53
        Rand3/Dir3, 53
        Rand4/Dir2, 54
        Rand4/Dir3, 54
        Rand5/Dir4, 55
strategy, 4, 29, **41**, 145
    analysis, **71**, 146
    antecedent, 41–42
    behavior, 43, **64**
        animal, 106
        social, 106
        objective vision, 83
    classical, 4
    control, 76
    correct choice, 4
    dynamics, 72
    examples, 47–63
    exploitation function, 70
    exploration function, 70
    four groups, 4, 42–46, *see* group
    number of individuals, 73
    self-adaptation, 6, **71**
    semi-directed, 4
    social, 106
    tests, 63
    type, 4, **43**
        influence, 76
    typical, 29
    unique conception, 4, 44–46
    unique formula, 4, 29, **42**, 71
strength pareto EA, 7
subpopulation, 6
summarizing scheme of DE, 7
support vector machines, 78, 121–123,
    148
    robustness, 123
SVM, *see* support vector machines
Swann, 101
swarm of particles, 104

temperature, 115
terminal condition, *see* stopping
    condition
test function, 14, 167, *see* standard test
    suite
  Ackley's function, 163
  quartic function, 162
  Rastrigin's function, 163
  Rosenbrock's function, 159
  rotated ellipsoid function, 164
  Shekel's function, 162
  sphere function, 159
  step function, 160
testbeds, *see* test function
testing, 63
Thomsen, 6
topographical principle, 6
traditional methods of optimization, 8
training set, 122, **123**
transversal DE, 94–97
  asynchronous parallelization, 98
transversal step, 94
trial individual, 17, 18, **29**, 71

trigonometric scheme, 42
Tvirdik, 6
two-additive measure, 135
two-array DE, 92
  synchronous parallelization, 98
typical evolutionary algorithm, 27

unconscious thinking imitation, 106
unconstrained problem, 28
unique conception, 4, 44–46
unique formula, 4, 29, **42**, 71, 145, *see*
    differentiation
universal algorithm, 101
universalization, **92**, 147
  example (FS), 95

value-to-reach, 29
variation, **26**, 30

Xie, 6

Zaharie, 6
Zelinka, 4, **32**
Zhang, 6

Printed in the USA